UNDERGRADUATE CHEMISTRY EDUCATION

A WORKSHOP SUMMARY

Keegan Sawyer and Joe Alper, *Rapporteurs*

Chemical Sciences Roundtable

Board on Chemical Sciences and Technology

Division on Earth and Life Studies

NATIONAL RESEARCH COUNCIL
OF THE NATIONAL ACADEMIES

THE NATIONAL ACADEMIES PRESS
Washington, D.C.
www.nap.edu

THE NATIONAL ACADEMIES PRESS **500 Fifth Street, NW** **Washington, DC 20001**

NOTICE: The project that is the subject of this report was approved by the Governing Board of the National Research Council, whose members are drawn from the councils of the National Academy of Sciences, the National Academy of Engineering, and the Institute of Medicine. The members of the committee responsible for the report were chosen for their special competences and with regard for appropriate balance.

This study was supported by the U.S. Department of Energy under Grant DE-FG02-07ER15872, the National Institutes of Health under Contract HHSN263201200074I (Task Order 25), and the National Science Foundation under Grant CHE-1231459.

International Standard Book Number-13: 978-0-309-29586-4
International Standard Book Number-10: 0-309-29586-6

Additional copies of this report are available from the National Academies Press, 500 Fifth Street, NW, Keck 360, Washington, DC 20001; (800) 624-6242 or (202) 334-3313; http://www.nap.edu.

Printed in the United States of America

THE NATIONAL ACADEMIES

Advisers to the Nation on Science, Engineering, and Medicine

The **National Academy of Sciences** is a private, nonprofit, self-perpetuating society of distinguished scholars engaged in scientific and engineering research, dedicated to the furtherance of science and technology and to their use for the general welfare. Upon the authority of the charter granted to it by the Congress in 1863, the Academy has a mandate that requires it to advise the federal government on scientific and technical matters. Dr. Ralph J. Cicerone is president of the National Academy of Sciences.

The **National Academy of Engineering** was established in 1964, under the charter of the National Academy of Sciences, as a parallel organization of outstanding engineers. It is autonomous in its administration and in the selection of its members, sharing with the National Academy of Sciences the responsibility for advising the federal government. The National Academy of Engineering also sponsors engineering programs aimed at meeting national needs, encourages education and research, and recognizes the superior achievements of engineers. Dr. C. D. Mote, Jr., is president of the National Academy of Engineering.

The **Institute of Medicine** was established in 1970 by the National Academy of Sciences to secure the services of eminent members of appropriate professions in the examination of policy matters pertaining to the health of the public. The Institute acts under the responsibility given to the National Academy of Sciences by its congressional charter to be an adviser to the federal government and, upon its own initiative, to identify issues of medical care, research, and education. Dr. Harvey V. Fineberg is president of the Institute of Medicine.

The **National Research Council** was organized by the National Academy of Sciences in 1916 to associate the broad community of science and technology with the Academy's purposes of furthering knowledge and advising the federal government. Functioning in accordance with general policies determined by the Academy, the Council has become the principal operating agency of both the National Academy of Sciences and the National Academy of Engineering in providing services to the government, the public, and the scientific and engineering communities. The Council is administered jointly by both Academies and the Institute of Medicine. Dr. Ralph J. Cicerone and Dr. C. D. Mote, Jr., are chair and vice chair, respectively, of the National Research Council.

www.national-academies.org

CHEMICAL SCIENCES ROUNDTABLE

CO-CHAIRS

WILLIAM F. CARROLL, JR., Occidental Chemical Corporation, Dallas, Texas
JENNIFER S. CURTIS, University of Florida

MEMBERS

MICHAEL R. BERMAN, Air Force Office of Scientific Research
CAROLE BEWLEY, National Institute of Diabetes and Digestive and Kidney Diseases
DONNA G. BLACKMOND, Scripps Research Institute
PAUL BRYAN, University of California, Berkeley
EMILIO BUNEL,* Argonne National Laboratory
ALLISON CAMPBELL, Pacific Northwest National Laboratory
A.WELFORD CASTLEMAN, JR., Pennsylvania State University
RICHARD R. CAVANAGH, National Institute of Standards and Technology
JOAN FRYE, White House Office of Science and Technology Policy
MIGUEL GARCIA-GARIBAY,* University of California, Los Angeles
JACK KAYE, National Aeronautics and Space Administration
JOHN KOZARICH, ActivX Biosciences, Inc.
LUIS E. MARTINEZ, Rollins College
KENNETH G. MOLOY, DuPont Company Experimental Station
ROBERT PEOPLES, Carpet America Recovery Effort
TANJA PIETRASS, National Science Foundation
MICHAEL E. ROGERS, National Institute of General Medical Sciences
ERIC ROHLFING, U.S. Department of Energy
JAMES M. SOLYST, ENVIRON International Corporation
KATHLEEN J. STEBE, University of Pennsylvania
PATRICIA A. THIEL,* Ames Laboratory and Iowa State University

NATIONAL RESEARCH COUNCIL STAFF

DOROTHY ZOLANDZ, Director
KATHERINE BOWMAN, Senior Program Officer
KATHRYN HUGHES, Senior Program Officer
DOUGLAS FRIEDMAN, Program Officer
KEEGAN SAWYER, Program Officer
ELIZABETH FINKELMAN, Administrative Assistant
SAYYEDA AHMED, Senior Program Assistant

JOE ALPER, Consulting Science Writer

---

* These members of the Chemical Sciences Roundtable served as members of the planning committee of the Workshop on Undergraduate Chemistry Education, but were not involved in the writing of this workshop summary.

Preface

The Chemical Sciences Roundtable (CSR) was established in 1997 by the National Research Council. It provides a science-oriented apolitical forum for leaders in the chemical sciences to discuss chemistry-related issues affecting government, industry, and universities. Organized by the National Research Council's Board on Chemical Sciences and Technology, the CSR aims to strengthen the chemical sciences by fostering communication among the people and organizations—spanning industry, government, universities, and professional associations—involved with the chemical enterprise. One way it does this is by organizing workshops that address issues in chemical science and technology that require national or more widespread attention.

On May 22-23, 2013, the CSR held a 1.5-day workshop on undergraduate chemistry education that focused on identifying potential drivers for change, barriers to curricular modifications, and new results from large-scale innovations with special emphasis on those that are transferable, widely applicable, and/or proven successful. The workshop featured both formal presentations and panel discussions among participants from academia, industry, and funding organizations. The workshop program consisted of three themes:

- Drivers of change in science, technology, engineering, and mathematics (STEM) education;
- Innovations in chemistry education; and
- Challenges and opportunities in chemistry education reform.

The workshop was intended to provide participants from a spectrum of the chemistry and chemistry education communities with an introduction to some of the work being done in this area, to stimulate further discussions, and to serve as a complement to other forums conducted by organizations such as the American Chemical Society, the Biennial Conference on Chemical Education, Gordon Research Conferences, and studies on undergraduate education conducted within the National Research Council. The Statement of Task for the workshop organizing committee is provided in Appendix A.

This document summarizes the presentations and discussions that took place at the workshop. In accordance with the policies of the CSR, the workshop did not attempt to establish any conclusions or recommendations about needs and future directions, focusing instead on issues identified by the speakers and workshop participants. In addition, the organizing committee's role was limited to planning the workshop. The workshop summary has been prepared by the workshop rapporteurs Keegan Sawyer and Joe Alper as a factual summary of what occurred at the workshop.

Acknowledgment of Reviewers

This workshop summary has been reviewed in draft form by persons chosen for their diverse perspectives and technical expertise in accordance with procedures approved by the National Research Council's Report Review Committee. The purposes of this review are to provide candid and critical comments that will assist the institution in making the published summary as sound as possible and to ensure that the summary meets institutional standards of objectivity, evidence, and responsiveness to the study charge. The review comments and draft manuscript remain confidential to protect the integrity of the deliberative process. We wish to thank the following for their participation in the review of this summary:

Diana Glick, Georgetown University
John Kozarich, ActiveX Biosciences, Inc.
David K. Lewis, Connecticut College (retired)
Marcy H. Towns, Purdue University

Although the reviewers listed above have provided many constructive comments and suggestions, they were not asked to endorse, nor did they see, the final draft of the workshop summary before its release. The review of this summary was overseen by **Edwin P. Przybylowicz, Eastman Kodak Company (retired)**. Appointed by the National Research Council, he was responsible for making certain that an independent examination of this summary was carried out in accordance with institutional procedures and that all review comments were carefully considered. Responsibility for the final content of this summary rests entirely with the authors and the National Research Council.

Contents

APPENDIXES

Acronyms

AAAS	American Association for the Advancement of Science
AAMC	Association of American Medical Colleges
ACS	American Chemical Society
BOSE	Board on Science Education
CPT	ACS Committee on Professional Training
CSR	Chemical Sciences Roundtable
DBER	discipline-based education research
GDP	gross domestic product
HHMI	Howard Hughes Medical Institute
iCons	Integrated Concentration in Science program
MCAT	Medical College Admission Test
MIT	Massachusetts Institute of Technology
MOOC	massive open online course
NMR	nuclear magnetic resonance spectroscopy
NRC	National Research Council
NSF	National Science Foundation
PCAST	President's Council of Advisors on Science and Technology
STEM	science, technology, engineering, and math
WIDER	Widening Implementation and Demonstration of Evidence-Based Reforms

1

Introduction and Overview

"It is clear that there is an enormous amount of activity in undergraduate chemistry education that is accelerating and intensifying."
Patricia Thiel

Undergraduate coursework in chemistry is a requirement for many university degree programs outside of the disciplinary fields of chemistry and biochemistry. Students hoping to pursue careers as doctors, dentists, biologists, chemical engineers, and environmental scientists, among other professions, are often required to take an introductory general chemistry course, if not also introductory courses in organic chemistry and biochemistry. As a result, effective science education is a topic of perennial interest to the chemistry community.

An upcoming change in the Medical College Admission Test (MCAT)[1] requirements is driving a recent increase in interest in the teaching of undergraduate chemistry (Brenner and Ringe 2012). New MCAT requirements may result in a change in the structure of chemistry as it is taught for pre-med students. When learning about some of the issues related to the MCAT modification, the National Research Council's (NRC's) Chemical Sciences Roundtable (CSR) felt it important to take the opportunity to examine some of the fundamental concerns and developments in the teaching of undergraduate chemistry.

On May 22-23, 2013, the CSR convened a public workshop, Undergraduate Chemistry Education, in Washington, D.C. The workshop explored drivers of science education reform and innovative approaches being implemented within chemistry departments to respond to some of these drivers. Workshop speakers described a variety of metrics and assessment tools for both drivers and innovations. Workshop discussions also explored barriers, opportunities, and realities of implementing reforms and modifications in today's chemistry curriculum.

In her introductory remarks at the workshop, organizing committee member Patricia Thiel of Iowa State University noted the enormous amount of activity in the field of science, technology, engineering, and mathematics education in general and in chemistry education in particular. Given this observation, the CSR aimed to hold an event that would be "valuable, fresh, and unique," said Thiel. Several recent publications informed workshop planning discussions, including the NRC Board on Science Education *Discipline-Based Education Research: Understanding and Improving Learning in Undergraduate Science and Engineering* (DBER report; NRC 2012); The President's Council of Advisors on Science and Technology report *Engage to Excel: Producing One Million Additional College Graduates with Degrees in Science, Technology, Engineering, and Mathematics* (PCAST 2012a); and a special issue publication of the journal *Science*, "Grand Challenges in Science Education" (McNutt 2013).

After considerable discussion and research, the workshop organizers decided to provide a forum focused on drivers of change, examples of educational innovation, and challenges and opportunities presented by chemistry education reforms. The focus on drivers of change was intended to raise awareness about some of the reasons why education reforms are being implemented and to illustrate that the motivation for change helps to define the metrics for measuring success. In planning the sessions devoted to educational innovation, the organizing committee decided to emphasize methods that can be used in large-scale (high-enrollment) situations such as the organic chemistry classes taught to nonchemistry majors at major universities. Some teaching methods are impractical with large groups even when they are "wonderfully successful" with smaller groups, Thiel explained. Therefore, she said, the workshop organizers decided to address the former

[1] The Association of American Medical Colleges (AAMC) oversees MCAT development and implementation. More information can be found on the AAMC webpage, MCAT2015 Exam for Students, https://www.aamc.org/students/applying/mcat/mcat2015/.

situation, because it affects large numbers of students and because larger courses are becoming increasingly common.

Thiel also noted the timeliness of this workshop given the announcement by the National Science Foundation (NSF) of a new program, Widening Implementation and Demonstration of Evidence-Based Reforms (WIDER),[2] which was released in the weeks leading up to the workshop. She acknowledged, too, that "there are different perspectives and controversies about almost every aspect of chemistry education. This meeting is not meant to cover every topic or every viewpoint or to represent every constituency but rather it is designed to help stimulate awareness and discussion."

ORGANIZATION OF THE WORKSHOP SUMMARY

This summary is organized into five chapters that are aligned with the major themes and goals of the workshop. Chapter 2 summarizes discussions on the drivers of change and the metrics used to identify the need for change in undergraduate chemistry education. The chapter begins with a broad look at the state of science in the United States and ends with drivers and lessons learned specific to chemistry education.

Chapter 3 describes innovative approaches to education reform, including key components and barriers to transforming large-scale undergraduate chemistry courses. Throughout the chapter, approaches and challenges with assessing the effectiveness of reforms is also discussed.

Chapter 4 describes the perspectives of four industry panelists on the state of undergraduate chemistry education and whether there is a need for change.

The final chapter recaps the final workshop panel discussion of five chemistry department chairs. The panel offered their insights and impressions on the state of undergraduate chemistry education, the types of innovations presented during the course of the workshop, and barriers encountered in trying to introduce novel instructional methods into the chemistry curricula at their institutions.

Although not comprehensive, this summary provides the readers with the key topics addressed during the workshop:

- Drivers of and barriers to change in chemistry education,
- Innovative course design for large-enrollment chemistry courses,
- Assessment tools needed to better evaluate the effect of novel course designs on chemistry learning,
- Industry and academic perspectives on the need for undergraduate chemistry education reform, and
- Potential next steps to more broadly disseminate innovative and effective chemistry course designs.

This publication is a factual summary of the presentations and discussions at the workshop. The views contained in the summary are those of the individual workshop participants and do not necessarily represent the views of all the workshop participants, the organizing committee, or the National Research Council. The summary does not contain any findings or recommendations about needs and future directions, but focuses instead on issues identified by the speakers and workshop participants.

[2] The WIDER program is overseen by NSF's Directorate for Education and Human Resources, Division of Undergraduate Education; http://www.nsf.gov/funding/pgm_summ.jsp?pims_id=504889.

2

Drivers and Metrics

"We all realize if the public were chemically educated, the world could be a much better place. And yet, we do not do as much as we could to address students in the life sciences, in the humanities, and [other disciplines]."
Miguel Garcia-Garibay

"Is there something wrong with chemistry education?" asked session chair Miguel Garcia-Garibay of the University of California, Los Angeles. "Are there things that need to be changed or are there simply opportunities to adopt new technologies, new skill sets from undergraduate students that perhaps could help us modify things? Is the need for more STEM professionals sufficient to make us rethink how we address undergraduate chemical education?" Garcia-Garibay asked these questions to start a discussion aimed at laying out the logic underpinning efforts to reform the way chemistry is taught to undergraduates, to science, technology, engineering, and math (STEM) majors as well as those in disciplines that use chemistry as an essential component of their skill set, such as pre-med students.

This chapter summarizes the presentations of five speakers at the workshop that addressed various aspects of why there might be a need to reform chemistry education and the ensuing open discussion. Alexandra Killewald of Harvard University discussed whether American science is in decline. Next, S. James Gates, Jr., of the University of Maryland and a member of the President's Council of Advisors on Science and Technology (PCAST), provided PCAST's perspective on the needs for STEM education and a STEM-educated workforce. Anne McCoy, of The Ohio State University and Chair of the American Chemical Society's (ACS's) Committee on Professional Training (CPT), described the role of the ACS Guidelines for Bachelor's Degree Programs in setting standards for undergraduate chemistry education. The potential impact of the Medical College Admission Test (MCAT) revisions on undergraduate chemistry education, one of the catalysts for this workshop, was discussed by Joel Shulman, of the University of Cincinnati and a member of the CPT. Last, Susan Hixson, who until her retirement in 2012 served as a program director in National Science Foundation's (NSF's) Division of Undergraduate Education, concluded the presentations with some lessons learned from NSF's experiences in undergraduate chemistry education.

IS AMERICAN SCIENCE IN DECLINE?

"Sometimes I think we are so focused on thinking about what is wrong with American science we do not take a step back to think about the fact the United States is actually the undisputed leader of contemporary world science in a way that is unprecedented in history," said Alexandra Killewald. Killewald cited statistics showing that the United States accounts for 40 percent of global research and development spending, 38 percent of new patented technology, and 45 percent of the Nobel Prizes in physics, chemistry, and physiology and medicine. Over one-third of scientific publications worldwide come from U.S. researchers, almost half of all citations are to papers written by U.S. authors, and nearly two-thirds of the papers published in highly cited journals come from U.S. laboratories. In addition, 15 of the world's top 20 universities are located the United States. "The influence of the U.S. on global science is enormous," Killewald emphasized.

If, as the statistics suggest, the United States is a global leader in science, why is there worry about the state of American science? Killewald and her colleague Yu Xie, of the University of Michigan, coauthored a book, *Is American Science in Decline?* (Xie and Killewald 2012), that takes a look at this issue. Killewald and Xie termed the position that society should be concerned about the state of American science as the "alarmist view." Killewald credited the alarmist view to the National Research Council's report *Rising Above the Gathering Storm* (NRC 2007). The NRC report raised the prospect of an impending shortage of U.S. scientists, which could affect American economic competiveness, said Killewald. She characterized the NRC report as "one of the most significant reports for U.S. science policy in recent history," noting that it led quickly to more than two dozen bills aimed at strengthening American science and the creation of a number of task forces to investigate this concern. Addressing whether there is evidence for the alarmist view, Killewald said, "to some extent, the answer is certainly yes," and she cited three main factors. First, the length of time between a Ph.D. program and the first independent science job is increasing. The number of years it takes to complete a Ph.D., the number of postdoctoral positions that emerging scientists need before securing their first job, and the number of postdoctoral fellows are all on the rise, explained Killewald. Second, there are unfavorable labor market outcomes for scientists. New scientists' wages are falling relative to the wages of other similarly trained professions, particularly lawyers and doctors. "Falling relative financial rewards might be one reason" why students might consider alternative careers to science, Killewald noted.

A third factor, one that Killewald says receives the most attention, is the idea that international competition, especially from continental East Asia, is threatening the dominant position of U.S. science. The average annual growth rate in output of science and engineering publications of eastern Asian countries far exceeds that of the United States, Europe, and Japan (see Table 2-1). Killewald stated that international

TABLE 2-1 Average Annual Growth Rate (%) in Science and Engineering Article Output

	United States	EU-15	Japan	East Asia-4
Biology				
1988-1992	1.7	6.4	4.6	17.7
1992-2003	1.1	4.1	3.9	16.0
Chemistry				
1988-1992	4.2	5.7	6.6	33.3
1992-2003	1.2	2.3	2.4	16.1
Physics				
1988-1992	5.1	10.6	10.9	19.7
1992-2003	0.3	3.4	4.4	14.3
Mathematics				
1988-1992	-2.0	3.2	-8.1	18.1
1992-2003	1.4	6.7	8.0	14.2

SOURCE: Harvard University Press.

growth in science is not "a prediction of doomsday," but an indication that the gap between the United States and other countries participating in global science" is narrowing. In terms of academic performance, schoolchildren in countries with economic resources similar to those of the United States, such as Hong Kong and Singapore, score substantially higher in math and science than those in the United States on a gross domestic product (GDP) per-capita basis. "This is the kind of result I see most commonly in the popular media," she said. However, Killewald maintained that "this picture is not one of failure to perform by the U.S., but it is a picture of average performance, and we might think we should do better than that."

Another area of concern is whether the United States relies too heavily on immigrant scientists. Killewald cited statistics showing that the physical sciences relative to other subfields have long had a slightly higher reliance on foreign-born scientists and continue to do so. In fact, the percentage of native-born Americans going into the physical sciences has declined steadily since 1960. As far as the student population is concerned, the fraction of foreign-born bachelor's degree students in science is only about 6 percent and that number has been steady since the late 1970s. It is only at the level of graduate degrees that there is an increase in the number of foreign-born students.

Despite the evidence in favor of the alarmist view, Killewald said there are "some real sparks of strength in the U.S. scientific picture." For example, the American scientific labor force is growing as a share of the total workforce. Also, in surveys of the general public, "scientist" continues to be regarded as a high-prestige occupation. In fact, the American public continues to express confidence in the leaders of the scientific community and to endorse public funding for basic scientific research. Academically, U.S. schoolchildren's scores on standardized tests in math and science are rising, and more U.S. students are completing advanced coursework. Killewald said that an increasing number of high school students are taking and passing Advanced Placement tests in science and math and an increasing number are taking calculus in high school.

There has also been no decline in the pursuit of scientific higher education over the past 40 years. Citing data from the National Center for Education Statistics (NCES 1972, 1980, 1988), Killewald noted that the percentage of students receiving bachelor's degrees who are in the top quartile of math achievement in high school has risen substantially over the past 40 years, with a nearly 50 percent increase among women getting bachelor's degrees (see Table 2-2). What has not changed much over that time is the percentage of men and women receiving science-related bachelor's degrees—nearly a third of men and approximately 13 percent of women. However, the percentage of students getting bachelor's degrees with a physical science major has fallen by over 50 percent for both women and men. The physical

TABLE 2-2 Bachelor's Degree and Science Major Attainment

	Male			Female		
	1972	1982	1992	1972	1982	1992
Bachelor's degree (%)	27.8	30.7	30.5	23.9	29.8	36.9
Among top 25% in math achievement	54.5	61.2	64.3	53.5	70.4	75.9
Science major given bachelor's degree (%)	28.7	31.4	28.3	10.2	13.7	13.2
Among top 25% in math achievement	36.9	41.5	38.8	15.7	20.9	19.3
Physical science	7.4	3.4	3.1	3.6	1.3	1.6
Life science	9.6	5.0	8.1	4.6	5.3	8.3

SOURCE: Harvard University Press.

sciences are losing some ground to engineering for males and the life sciences for females. Killewald said that the same trend appears to be holding true for graduate degrees, both in science overall and the physical sciences specifically.

Returning to her original question—Is American science in decline?—Killewald said she and Xiu contend the answer is a qualified no. "We think the evidence of health in American science generally outweighs the concerns." She acknowledged that "the question of whether you think American science is doing well or not depends on your point of comparison." From an international perspective, it could be said that America's *leadership* in science may soon be challenged. "It is easy to see looking in your rearview mirror that other folks are catching up fast." From a historical perspective, U.S.-based science is not in decline, but rather is "doing as well as or better than before in terms of our own performance." As a final thought, Killewald emphasized that it is important to remember that there are collaboration benefits arising from globalization in addition to competition costs. "The rise in science in other countries brings new perspectives to the scientific enterprise" that can result in scientific advancement and benefit the American people.

Matthew Tarr, from the University of New Orleans, commented that he has seen a dramatic increase in the number of chemistry majors and students taking general chemistry courses over the past 3 years and asked if Killewald had more recent data on national trends. She replied that data from the cohort of students who graduated from high school in 2002 were not yet available when she and Xie wrote their book, but that this was likely to be the case given the rise in students going into the medical sciences. She added that impacts of the Great Recession are likely to include students placing an increased emphasis on taking courses that will lead to jobs, and she expected that fact to increase enrollment in scientific courses.

David Harwell from the ACS commented that while the supply of graduates with science degrees may have remained constant or increased in some areas, the demand side of the equation does not look as good. Research by the ACS indicates that innovation is down compared with that in other countries, unemployment rates among chemists are up, and salaries have fallen in inflation-adjusted terms. Killewald responded that these data support the idea that the problem is one of oversupply, not a shortage in some fields and particularly in academia.

A PCAST PERSPECTIVE ON STEM EDUCATION IN THE NEW MILLENNIUM

S. James Gates, Jr., described the role of PCAST, a civilian advisory group that makes science policy recommendations to the President, and PCAST's activities and positions on science education and workforce. During the Obama Administration, PCAST has produced several reports focused on STEM education and workforce: *Prepare and Inspire: K-12 Education in Science, Technology, Engineering, and Math (STEM) for America's Future* (PCAST 2010) and *Engage to Excel: Producing One Million Additional College Gradu-*

ates with Degrees in Science, Technology, Engineering, and Mathematics (PCAST 2012a). Referring to Killewald's presentation, Gates said that he and his PCAST colleagues agree with the alarmist view, given their prospective, rather than retrospective, view of the health of the U.S. science and technology enterprise. Of particular concern, he said, is the decreased amount the country is investing in science, noting U.S. investment in science has now fallen to around 3 percent of GDP. Compared with the rest of the world, this figure is "middling," said Gates. Even more concerning to PCAST is that the balance between high-risk/high-reward funding and low-risk funding is suboptimal for the nation's future. In response to these science funding concerns, PCAST issued a report in November 2012, *Transformation and Opportunity: The Future of the U.S. Research Enterprise* (PCAST 2012b), that laid out a set of metrics that funding agencies might begin to use as they think about how to fund research.

Gates acknowledged that there are number of different ways to examine U.S. performance in STEM relative to that of its global competitors. Some metrics suggest that the country is doing fine. However, one signal that the nation is underperforming emerges from evaluations of what PCAST calls the STEM-capable workforce. The STEM-capable workforce ranges from STEM professionals in STEM jobs, such as academic research, to STEM-trained professionals in non-STEM jobs that require STEM skills, such as health care or advanced manufacturing (PCAST 2012a). The latter type of jobs "are going unfilled today in the aftermath of the Great Recession, and it is the lack of Americans with the STEM training to fill these jobs that concerns PCAST," Gates said. The STEM skill set is growing in value in the United States, but employers are having difficulty finding people with the adequate expertise for these positions.

PCAST's 2012 *Transformation and Opportunity* report focuses on how to make sure that the benefits of STEM education extend to the entire American economy to create the possibility that the American Dream will be extended to another generation, said Gates. The report is not about how to "reproduce" academic researchers more efficiently. Concerning the issue of underperformance, Gates discussed trends in the attainment of college degrees. Among 25- to 64-year-olds, the United States ranked third, according to 2008 data, behind Japan and Canada in terms of percentage of the population with college degrees. But, the United States dropped to ninth among 25- to 34-year-olds. "The current youngest generations of Americans in the workforce are technically less well educated" than the generation preceding them, said Gates. "This is the first time in over 100 years this statement could be made." These data are worrisome, he added, because the nation's economy has entered a period when the wage premium associated with a college degree is increasing rather dramatically.

Using data from the Bureau of Labor Statistics, PCAST found that between 2008 and 2018, STEM occupations will increase from 5.0 percent of total jobs in the United States to 5.3 percent, an increase equivalent to one million jobs from growth alone. In addition, over one million jobs that exist currently will need replacement employees to account for turnover in the workforce. In other words, the projected number of life, physical, and social science technician job openings will far exceed the number of STEM-trained individuals to fill those positions. PCAST also found that the gap between supply and demand will vary by discipline. For example, there are some signs that the nation may be overproducing people trained in the biological sciences and underproducing in computer sciences.

In its studies, PCAST found that retention and diversity problems in STEM undergraduate education are significant, said Gates. Fewer than 40 percent of students who enter college intending to major in a STEM field complete a STEM degree (PCAST 2012a). High-performing students frequently cite uninspiring introductory courses as a reason for changing majors. PCAST found that low-performing students with a high interest and aptitude in STEM face difficulty in introductory courses resulting from insufficient math preparation and help. Many of the low-performing students cite an unwelcoming atmosphere from faculty teaching STEM courses as their reason for switching majors. Women and members of minority groups now constitute approximately 75 percent of college students, but only 45 percent earn STEM degrees. Women and minorities are leaving STEM majors at higher rates than other groups of students, said Gates, thus constituting an expanding pool of untapped talent.

The question of how to diversify STEM pathways is a big one. The economy is entering a period in which people will not have one career for 40 years but rather will need a broad set of STEM skills that will allow them to adapt to new opportunities and even undergo retraining at some point in their working lives. Gates contended that this shift will require that the current pipeline model of STEM education change to accommodate multiple on-ramps and off-ramps for people to get into and out of STEM training.

To address these STEM education and workforce challenges, PCAST made four recommendations in the *Engaged to Excel* report. The first, which Gates predicted would be a challenge for today's faculty members, is to catalyze widespread adoption of empirically validated teaching practices, that is, evidence-based learning. PCAST's goal is that successful programs should be expanded to reach 10 to 20 percent of the nation's 230,000 STEM faculty over the next 5 years, by providing training to existing faculty but also by requiring that all graduate students and postdoctoral fellows supported by federal training grants will receive instruction in modern, evidence-based teaching methods. PCAST acknowledged in its report that making this transition has a cost and recommended that the federal government provide \$10 million to \$15 million a year for the next 5 years to fund this effort. PCAST also called for the development of metrics by which institutions can gauge their progress toward excellence in STEM education.

The second recommendation calls for undergraduate STEM programs to replace standard laboratory courses with discovery-based research courses. This recommendation comes out of the premise that students who engage in research early in college are more likely to persist in STEM majors. Gates pointed to the Freshman Research Initiative at the University of Texas[1] as an example of a program demonstrating the value of early research opportunities for retaining students in STEM degree programs. PCAST noted that research universities and small colleges should form collaborations to provide all students with access to research opportunities. PCAST's third recommendation says the nation should launch an experiment in postsecondary mathematics education to address the math preparation gap. This recommendation came from the fact that college-level skills in mathematics and computation are a gateway to other STEM fields but that nearly 60 percent of students enter college without the math skills needed for STEM majors, something that Gates personally finds appalling. Addressing this gap will provide access to great opportunities to the 14 percent of 12th-grade students who express interest in STEM fields but do not currently have the math skills to pursue those interests.

The final PCAST recommendation calls for the creation of partnerships among all stakeholders to diversify pathways to STEM careers. It is critical to engage all of the end users of STEM-trained individuals. This call will require efforts that must go beyond academia to be successful, said Gates.

In response to a question from Mark Cardillo of the Dreyfus Foundation about the role that online courses can play, Gates said that PCAST looked specifically at massive open online courses, also known as MOOCs, and supports leveraging information technology to improve the efficiency of teaching. Technology, however, is not going to replace teachers or professors, noted Gates. It can empower educators and radically change the environment in which they function. The key will be to figure out how to engage this technology in a way that leverages what individual teachers do to improve STEM education.

David Harwell of the ACS asked if PCAST had considered how to fill the need for people with associate- or certification-level training to fill jobs in fields such as chemical manufacturing and if there was any thought given to pushing the 14 percent of students who have a high interest in STEM careers but poor math skills toward programs that would fill those needs. Gates replied that he personally is not in favor of pushing students in any direction. "I want the students to be active agents in making choices," he said. "After all, that is the great thing about democracy and the type of economic system we have. You want to offer variegated choices to people so individuals will make the choice they see as best for themselves." Gates noted that the Obama Administration believes the way to address this issue is to upgrade the community college system, an approach that PCAST supports.

ROLE OF THE ACS GUIDELINES FOR BACHELOR'S DEGREE PROGRAMS

Ann McCoy discussed the role of the CPT and the ACS Guidelines for Bachelor's Degree Programs. ACS established the CPT in 1935 to assume responsibility for properly accrediting institutions wishing to grant undergraduate chemistry degrees. Today, the committee's goals are to promote and assist in the development of high standards of excellence in all aspects of postsecondary education, undertake studies important to the maintenance of these standards, and to collect and make available information about trends and developments in modern chemical education. In addition to establishing and administering the degree accrediting program, the CPT devotes a significant amount of time conducting surveys to understand current trends in areas related to the professional education of chemists. The committee also compiles the ACS Directory of Graduate Research and coordinates workshops and other activities that bring together members of the chemistry education community. The CPT is currently in the process of revising the bachelor's degree guidelines.

The CPT sets the ACS Guidelines for Bachelor's Degree Programs in chemistry. The CPT uses the guidelines as the basis for approving degree programs; currently, 669 programs are accredited under this process. The chairs of individual departments then certify students who meet the approved program curricula. She said that the approved programs benefit both the students who receive the certified bachelor's of science degrees and all other students taking classes in those departments because of the supportive infrastructure that must exist to become an approved program. In fact, while the number of students receiving certified bachelor's degrees has risen slightly since 1950, the number of overall chemistry degree graduates has more than tripled during the past six decades. McCoy noted that although about half of the students receiving certified degrees come from a small number of the institutions with the largest graduate programs, the guidelines serve to provide a level of uniformity in programs and standards of excellence that benefit all students, as well as the profession as a whole.

The guidelines include requirements for institutional involvement, faculty and staff numbers and their contact with students, and infrastructure, but McCoy focused her talk on the curriculum requirements in the guidelines. The guidelines are not designed to constrain programs by mandating a set curriculum, but to provide opportunities to gain the resources and infrastructure needed and guidance in terms

[1] See http://fri.cns.utexas.edu/.

of what it means to educate professional chemists, explained McCoy. Opportunities for undergraduate research are an essential component of the curriculum guidelines, along with student skill development and departmental self-assessment in terms of which aspects of a curriculum are working and which could use improvement. The overall philosophy of the guidelines is not to be overly prescriptive in terms of specific course requirements or laboratory experiences, but instead to provide a scaffold on which programs develop a curriculum that is appropriate for their students.

There are specific course requirements, but even those are fairly loose, said McCoy. Students need to take a foundation course in each of the five traditional areas of chemistry (organic, inorganic, biochemistry, analytical, and physical), and four in-depth courses based on that foundational experience. The guidelines do not detail the precise nature of those in-depth courses. Students also need a minimum of 400 laboratory hours after general chemistry, and those hours need to cover at least four of the five traditional areas. In addition, all nine of the required courses and the required labs must be offered annually, a requirement that can be challenging for smaller chemistry programs to meet but is deemed necessary to ensure that students can graduate in 4 years. McCoy noted that programs are encouraged to include contemporary topics in chemistry and to employ a variety of approaches in delivering this curriculum. The 2008 revisions of the guidelines placed a stronger emphasis on professional skills such as problem solving, using the chemical literature, laboratory safety, oral and written communications, working in teams, and ethics.

The CPT is currently in the process of revising the guidelines. McCoy expects the new guidelines to be adopted in 2014. The revision process began with a survey of approved programs on the impacts of the 2008 guidelines (results are accessible through the CPT website).[2] Overall, the survey indicated that curricular changes based on the 2008 guidelines were modest, likely reflecting the short time period since the 2008 guidelines were introduced and the additional fiscal stresses felt by departments since 2008. Approximately two-thirds of the programs had no trouble offering an approved curriculum, but 25 percent of the programs reported occasional difficulties. In response to this finding, the proposed revisions call for increasing the minimum faculty size from four to five individuals by 2025. McCoy noted that the CPT had proposed this same change for inclusion in the 2008 guidelines, but backed off in response to community pushback.

In January 2013, the CPT issued a white paper on possible guideline revisions with the goal of soliciting comments from as broad a swath of the community as possible.[3] One area in which the CPT received significant input concerned courses that are largely or exclusively offered online. A year ago, more than 10 percent of the programs that responded to the survey offered online general chemistry courses. Only a few percent also offered foundation and in-depth courses online. Over half of the surveyed programs felt that online courses were inappropriate, though about a quarter of the programs believed that the online venue could serve a role in providing introductory courses. A very small percentage of programs thought online courses were appropriate for meeting degree certification requirements. Fewer than 5 percent of programs offer online laboratory courses. While over half of the responding departments said that virtual laboratories were inappropriate, more than 40 percent thought virtual laboratories could serve a limited, supplementary role. In response to the survey, the CPT has proposed requiring programs to provide significant hands-on laboratory experiences prior to starting the foundational lab experience, explained McCoy.

Among the surveyed departments, there was near universal agreement that undergraduate research is a great experience for students. There was a strong consensus, McCoy iterated, that the guidelines should require an undergraduate research experience, but such a requirement would be difficult to implement, particularly by smaller programs. In thinking about what students would gain from this experience, the CPT concluded that it was not conducting research per se, but rather the opportunity to apply all of the skills and ideas they have gained as students to a personalized learning experience. In the end, the CPT proposed introducing a requirement for a "capstone experience." Capstone experiences—which could include research, a group problem-solving class, an internship, or mentored teaching, among other possibilities—would provide students with opportunities to synthesize the knowledge and skills they gained across the curriculum.

Another common issue raised by survey respondents is concern about removing the requirement for two semesters of both organic and physical chemistry. The CPT has made this compromise to introduce flexibility into the curriculum. McCoy explained that about 4 percent of the programs have introduced a one-semester integrated organic chemistry course and 1 percent reduced the physical chemistry requirement to one semester for at least one degree track. Only 1 to 3 percent of programs are considering making similar changes.

The proposed changes would also alter the guidelines' instrumentation requirements. Recognizing the importance and expense of gaining experience with nuclear magnetic resonance (NMR) techniques, the revised guidelines would allow programs to use an offsite NMR facility to fulfill this requirement. The guidelines would also require student expo-

[2] See http://www.acs.org/content/acs/en/about/governance/committees/training.html.

[3] See http://www.acs.org/content/dam/acsorg/about/governance/committees/training/guidelines-white-paper.pdf.

sure to at least one instrument in each grouping of optical atomic spectroscopy, optical molecular spectroscopy, mass spectrometry, electrochemistry, and chromatography/separations (McCoy and Darbeau 2013).

McCoy said that she sees the guidelines and the revision process as a community activity. She encouraged the workshop attendees to provide her and her colleagues Clark Landis and Joel Shulman, both presenters at the workshop, with comments, or to send comments to the ACS via e-mail at cpt@acs.org.

In response to a question about requiring a capstone experience rather than an actual research experience, McCoy said that the problem is not that faculty are not enthusiastic about providing a research experience, but that doing a good job for every student seeking a certified degree would be challenging. She noted one program where students work with industry chemists who pose challenges for the students to solve in teams. "That is almost as good, or maybe better, than doing more traditional undergraduate research projects," she said. On the other hand, undergraduate research also provides important mentoring opportunities for graduate students and McCoy said that finding the right opportunities for each student will be key. "Requiring undergraduate research of all certified majors seems to be something the community is very concerned about," she said.

One attendee asked McCoy if she was surprised at the strong negative response to online courses. She answered that the negative view could reflect the conservative nature of the chemistry community, or it may also reflect a misinterpretation of the survey question. McCoy explained that the survey question about online teaching did indicate the use of online instruction in combination with in-class teaching. She also added that she has seen some great opportunities for doing shared online instruction between multiple smaller institutions.

Responding to a question about how the courses offered at 2-year institutions fit into the guidelines, McCoy said that there is a separate set of guidelines for 2-year colleges that were adopted shortly after the 2008 guidelines (ACS 2008) were put into place. She also noted that there is a new 2-year college advisory board and the CPT has representation on that board. "It is an ongoing process but it is one I think we have made a lot of progress on in the last 3 or 4 years," she said.

CHEMISTRY AND THE PRE-MEDICAL CURRICULUM: IMPACT OF MCAT^{2015}

Joel Shulman of the University of Cincinnati and CPT member discussed potential impacts of MCAT changes on undergraduate chemistry. A report from the Association of American Medical Colleges (AAMC) and the Howard Hughes Medical Institute (HHMI), *Scientific Foundations for Future Physicians* (AAMC/HHMI 2009), advocates a new focus for both pre-medical education and medical school curricula on core competencies rather than on specific courses or disciplines. On the basis of the report's findings, AAMC is working to transform medical school admissions to keep pace with the changes in science and medical education with the ultimate goal of preparing a workforce that can better care for Americans' health, said Shulman. As part of that transformation, students wishing to apply to medical school will begin taking a revamped MCAT starting in 2015.

The MCAT^{2015} will consist of four sections and generate four scores, one of which will be on the chemical and physical foundations of biological systems, while another will cover the biological and biochemical foundations of living systems. Questions in these sections will require students to have an understanding of the principles that govern chemical interactions and how these reactions form the basis for a broader understanding of the molecular dynamics of living systems (Schwartzstein et al. 2013). They will test introductory-level organic and inorganic concepts—biochemistry concepts at the level taught in most first-semester biochemistry classes—and target basic research methods and statistical concepts described by many baccalaureate faculty as being important to success in introductory science courses (AAMC 2011). Shulman explained that the approximate distribution of questions in the section on the chemistry and physical foundations of biological systems will be 30 percent general chemistry, 25 percent organic chemistry, 15 percent first-semester biochemistry, 25 percent introductory physics, and 5 percent biology. He noted that this is not much different than the subject matter distribution of the current MCAT test with perhaps a little more biochemistry.

The effect that the new MCAT will have on undergraduate chemistry courses that pre-med students are required to take is unclear. MCAT^{2015} will assess a set of eight scientific competencies (the combination of skills, abilities, and knowledge needed to perform a specific task) as designed by the AAMC (see Figure 2-1), explained Schulman. Two of these competencies (highlighted in Figure 2-1) are related directly to chemistry or biochemistry. For example, Competency E4 will require students to demonstrate knowledge of basic principles of chemistry and some applications of those principles to the understanding of living systems.

What is the best way that chemistry departments ensure that pre-med students master these core competencies? Shulman suggested three possible approaches: (1) "apply the concepts of chemistry to biological principles in biology courses"; (2) "apply a biological context to chemical principles in chemistry courses"; or (3) do both 1 and 2. In his opinion, it makes sense to do both. Shulman noted, however, that general and organic chemistry should be making as many connections to biology as possible regardless of whether these subjects are taught together or separately.

The new approach to testing medical school applicants raises the question of whether the chemistry curriculum

Competency E1 *Apply quantitative reasoning and appropriate mathematics to describe or explain phenomena in the natural world.*	Competency E2 *Demonstrate understanding of the process of scientific inquiry, and explain how scientific knowledge is discovered and validated.*
Competency E3 *Demonstrate knowledge of basic physical principles and their applications to the understanding of living systems.*	Competency E4 *Demonstrate knowledge of basic principles of chemistry and some of their applications to the understanding of living systems.*
Competency E5 *Demonstrate knowledge of how biomolecules contribute to the structure and function of cells.*	Competency E6 *Apply understanding of principles of how molecular and cell assemblies, organs and organisms develop structure and carry out function.*
Competency E7 *Explain how organisms sense and control their internal environment and how they respond to external change.*	Competency E8 *Demonstrate an understanding of how the organizing principle of evolution by natural selection explains the diversity of life on earth.*

FIGURE 2-1 Competencies for entering medical students. SOURCE: Adapted from AAMS/HHMI (2009).

should change in response to MCAT^{2015}. Currently, most pre-med students complete five semesters of chemistry—one year each of general and organic chemistry and one semester of biochemistry. Shulman said that he sees no reason that a school or program has to make a change in its curriculum for pre-med students. However, there are opportunities and challenges for the chemistry community that are laid out in *Scientific Foundations for Future Physicians* (AAMC/HHMI 2009), Schulman said. He argued that a major opportunity is for the chemistry community to recognize that most freshman and sophomore chemistry students, and not just pre-med students, have a strong interest in biology-related curricula. Schulman reiterated throughout his talk that the chemistry community should consider introducing more biological examples into both general and organic chemistry, regardless of MCAT^{2015}.

It should be possible, Shulman continued, to take advantage of the flexibility in the ACS guidelines that McCoy described to reorganize chemistry curricula to emphasize the biological aspects of chemistry. He described several approaches that an ACS task force has identified for doing so. One approach would be to integrate biological examples into the traditional curriculum, and Shulman gave several examples of this. Enzymatic catalysis can be discussed when teaching about other catalytic processes, including the role of proximity within active sites and nonbonding interactions; peptide bonds and protein conformations as part of the study of carboxylic acids and amide bonds rather than as separate topics, usually at the end of the semester; and biologically relevant types of reactions, such as the Claisen condensation to form acetyl coenzyme A, or the formation of sulfates and phosphate bonds that are relevant to biological molecules.

Another approach to emphasize the biological aspects of chemistry is to create two different second-semester organic chemistry classes, one focused on bioorganic chemistry for pre-med and other biology-oriented students, and the other course emphasizing mechanism and synthesis for chemistry majors and chemical engineers. Shulman noted that Oberlin College has been using this construct successfully for 20 years, and though it requires the availability of teaching resources to offer two different second-semester organic chemistry courses, the ACS task force found that at least a few institutions are trying this approach.

Purdue University, with HHMI funding, has been developing what is being called the 1-2-1 approach, a 2-year curriculum for freshman and sophomore students. Each year consists of one semester of general chemistry, two semesters of organic chemistry, and one semester of biochemistry. In the 2-year curriculum, the general chemistry courses have a strong acid–base emphasis with connections to biochemistry. The organic chemistry courses emphasize reactions and mechanisms with biochemical analogies, while deemphasizing retrosynthesis and organometallic chemistry (Shulman 2013). The 1-2-1 approach assumes that students are adequately prepared before college so that one semester of freshman-level general chemistry is sufficient for success in the subsequent organic chemistry courses. Juniata College in Pennsylvania has used the 1-2-1 curriculum for years, with chemistry majors taking an additional year of organic chemistry as juniors.

A fourth possibility is an "organic chemistry first" approach. Freshman start with what Shulman called a "biologically flavored" organic chemistry course that introduces some general chemistry concepts intercalated with relevant biology. The modified organic chemistry course is then followed by either two semesters of mainstream chemistry or one mainstream chemistry course and one biochemistry course. The most radical approach, said Shulman, would replace the standard first 2 years of chemistry with a one-semester course on structure and properties and a three-semester sequence on reactivity. The College of St. Benedict & St. John in Minnesota has developed this more extreme approach and is making a significant effort to link these courses to the MCAT^{2015} core competency requirements. Shulman added that this curriculum is allowed by the ACS guidelines but has not yet been reviewed by the CPT for approval. Schulman noted that the College of St. Benedict & St. John approach "takes a lot of work and it takes a lot of coordination."

Commenting on the challenges of making curriculum changes to meet the demands of the new MCAT, Shulman said that smaller schools may have difficulty accommodating the chemistry requirements for all majors. He added that any type of curricular change takes buy-in from the faculty, coordination among departments, and the availability of appropriate texts. This last issue could be a particular problem, said Shulman. The CPT has discussed curriculum change with textbook publishers, but "they are not going to write textbooks until they know there is a large enough audience, and in many cases there will not be a large enough audience until there are textbooks," he observed. Other challenges include coordination between 2-year and 4-year colleges to ensure that transfer students can transition smoothly into a new curriculum, and the potential impact on the need for teaching-assistant support at large schools, particularly if a curriculum moves from a two-semester to a one-semester general chemistry sequence.

Shulman pointed out that there are still many unanswered questions. He asked, "Will medical schools have the ability and desire to adjust their admission requirements to do away with course requirements and reflect competencies almost completely?" Other questions are whether undergraduate programs will be motivated to map courses onto pre-medical competencies, and whether the new MCAT will successfully assess competencies with credibility and reliability. Schulman highlighted a commentary by Charles Brenner and Dagmar Ringe (2012) that was published in *ASMBM News* that recommended going to a 1-year-of-chemistry and 1-year-of-biochemistry curriculum for pre-med students, with the 1-2-1 curriculum as an intermediate step toward this curriculum. Shulman rejected that idea, asserting, "I do not think you can possibly do students a service by going to that model."

In closing, Schulman noted that there will be a series of commentaries in an upcoming issue of the *Journal of Chemical Education*, including one by Charles Brenner that will discuss the role of chemistry in the pre-med curriculum (Brenner 2013). He said that the bottom line is that the chemistry community needs to see how the MCAT is constructed and how it treats the intersection of content and skills in chemistry. "We need to figure out what our metrics ought to be so we know that any pedagogical changes made will be meeting the needs not only of the pre-medical students but all students."

William Tolman, from the University of Minnesota, questioned Shulman's statement concerning the biological interests of most first- and second-year chemistry students. At his institution, students have been "flocking away" from a biologically oriented class to the one that is less biologically oriented. Shulman responded that he had no statistical evidence, but that his conversations with organic chemistry faculty support this view.

David Harwell asked if it was also important to consider the competencies that the chemical industry needs, and not just those of medical schools, when thinking about redesigning curricula. "Should chemical educators be looking at more competencies as opposed to the courses we normally teach in preparation for graduate school?" he asked. Shulman supported this idea but noted that competency-based education is a challenge without good metrics to measure competencies accurately.

LESSONS LEARNED AT NSF

There are two homes for undergraduate chemistry education at NSF—the Division of Undergraduate Education in the Directorate for Education and Human Resources, and the Chemistry Division of the Directorate for Mathematics and Physical Sciences—explained Susan Hixson, who noted that her comments do not necessarily represent official views of the NSF. She emphasized that there is "a boatload of existing results on successful undergrad chemistry education interventions, including content and pedagogy." She noted, too, that active learning strategies have been perfected for the chemistry community and that there continues to be a significant research effort to better understand student learning in the context of undergraduate chemistry education research. Much of this research is published in the *Journal of Chemical Education*,[4] highlighted in the Chemistry Education Division sessions at the semiannual ACS national meetings and also highlighted in the biannual Gordon Research Conference Programs on chemistry education and chemistry teaching. There have also been dozens if not hundreds of reports from policy groups, professional organizations, and other

[4] See http://pubs.acs.org/journal/jceda8.

interested parties on STEM education, and the ACS will be publishing a book that tracks many major research efforts.

In making some general observations, she noted that there are blurred conversations on education in terms of grade level, the type of undergraduate student, and the goals of transformation efforts. For example, it is important to know whether the discussion is about "generating chemistry majors who will go to graduate school," "bachelor- or two-year college-level majors who are going to have industrial jobs," or "producing K through 12 teachers," she said. Hixson added that there are two factors that are unique to chemistry when it comes to discussing drivers of reform for teaching undergraduate chemistry: (1) first- and second-year classes are typically full because of the demands from majors other than chemistry or chemical engineering, and (2) the chemical profession actually worries when chemistry undergraduates are not finding jobs.

Developing and implementing reforms takes a great deal of effort sustained across a long, complex process, said Hixson. Introducing active learning techniques, for example, depends absolutely on having a committed, obsessive faculty member. "If there is going to be a major kind of change in how a course is taught, you have to assume it will take a decade or more for that reform to hit a national level," she explained. In that regard, the chemistry community was fortunate to have the backing of NSF's Chemistry Initiative 1994-1999 that not only introduced new ways of teaching undergraduate chemistry but also generated a huge cadre of faculty who were familiar with undergraduate education in chemistry and led to the development of a much larger chemistry education research field, she added.

One question that arises during any reform effort is why a project does not persist at a developer's institution. One reason, she said, is that it did not work. Another is that the reform effort was led by a single faculty member who lost interest or left the institution. Changes in technology platforms and institutional changes, such as budget constraints or even the appointment of a new department chair or dean, can also cut short the life of a reform effort.

The failure of a successful effort to travel from one institution to another is because curriculum developers often forget to involve faculty from other institutions at the beginning of their projects. The result is a program that is idiosyncratic to the faculty at the home institution, Hixson explained. Developers also underestimate the sustained effort it takes to perfect and then disseminate a program. In addition to creating materials and pedagogy, and testing and revising them, a developer needs to assemble a group of colleagues who will speak at professional meetings and hold workshops for potential adapters, all of which requires funding, usually from sources outside of an institution. Cross-departmental projects are particularly challenging to develop and implement, Hixson explained, and require the sustained commitment at multiple institutional levels.

Hixson said funding agencies or foundations will often support educational reform efforts by funding scholarships or internships directly to students. While the motive is laudable, the benefits then travel with the student and often make little or no impact on the infrastructure at the host institution. In the same vein, programs that provide research opportunities for undergraduates have the same problem, and also are often limited to the "best and brightest" upper-level students and thus have little impact on expanding the pool of chemistry majors. There is also little information on whether research opportunities are effective at meeting their goals. In that regard, evaluating the effects of any reform effort is still challenging, Hixson noted. Too often, assessment is done too early in the life of a project or funding ends before evaluation is complete.

Hixson said that there were many missed opportunities in chemistry during her 20 years at NSF. One example, she said, is that while ACS has great national meetings, the Division of Chemistry Education has such a large program that its sessions almost always occur at a site separate from the rest of the ACS meeting. As a result, there is less cross-fertilization among faculty than might have been expected. In addition, most ACS journals do not accept education papers, again limiting cross-fertilization. New ACS presidents typically have some focus on education, but they could be better informed on the subject, she said. Although the CPT is known for its emphasis on chemistry content for chemistry majors, it has had little impact on the pedagogy for nonchemistry and nonscience majors. Another problem she pointed to was the fact that the ACS website only points to the society's own work in the field, in contrast to the American Physical Society's website, which points to major efforts throughout the field. The Gordon Research Conferences should be encouraged to include relevant education talks in their extensive offerings in chemistry. The Pittcon conference started doing this in the 1990s for analytical chemistry, she said in closing.

In response to a question from Matthew Tarr about incentives for faculty to participate in curriculum reform efforts, Hixson noted that this is the number one excuse she hears. She responded that while it is absolutely true that tenure decisions are based largely on research productivity, the tenure period typically lasts a mere 6 years, leaving decades for a faculty member to work on education issues. Hixson added, however, that she does not believe that the field suffers from a lack of successful interventions, but rather from not implementing the many effective ones that already exist.

DISCUSSION

The first issue raised during the open discussion period focused on how to link information learned in classrooms to real-world matters. James Anderson of Harvard University noted that students come to Harvard as masters of the standardized test and that it is a challenge to get them to start to think about systemic, integrated issues that include the human body but also include global energy concerns, climate change, and others. In that regard, Anderson cautioned against going too far toward accommodating changes in the chemistry curricula to meet the needs of the MCAT. Shulman agreed and said that a general failure of chemistry curricula is that they do not demonstrate much connection to any of these broader issues. Cardillo added that chemistry education should not overlook the nonscience major. He argued that chemistry curricula have an opportunity to educate the general public through the nonscience major by emphasizing the connections between chemistry and broad topics that grab the attention of this larger audience, the citizens of the nation. For science majors, Hixson said that while the goal should not be to turn every student into a chemistry major, the field needs to do a better job informing students about the career options available for people with STEM degrees.

Thomas Holme, from Iowa State University and director of ACS Exams Institute, pointed out that the 2013 freshman class will be the first cohort that has been subjected to nonstop standardized testing since fourth grade as a result of the No Child Left Behind law. He asked whether this is a concern. Killewald said that her understanding is that No Child Left Behind has improved math performance and that she would not anticipate a negative effect on the preparation of entering students.

Robert Peoples, of the Carpet America Recovery Effort, focused on the content of chemistry coursework in light of scientific advances. He noted that chemistry faculty members have been teaching the same chemistry content using the same techniques for the last 100 years despite significant advances in chemistry. Peoples believes that it is important to think carefully about curriculum content and teaching contexts, and so cautioned against cramming more information into the same courses. McCoy replied that CPT has been considering issues about content and context. CPT believes that it is important to design the ACS guidelines to be less prescriptive about content, and instead place greater emphasis on teaching methods that help students build an understanding of how chemistry works and the language of different areas of chemistry. McCoy emphasized that it is important for chemistry students to be able to think more about broader topics and communicate across fields. “At the end of the day, less may be more,” said McCoy.

Jody Wesemann, from the ACS Education Division, asked whether infrastructure development might be needed to better prepare students to meet an uncertain future. McCoy answered that chemistry departments need to develop a physical plant that has the flexibility to allow for all of the different types of teaching modalities and teaching styles, such as online access to material outside of the confines of a lecture hall. She also stated that teaching laboratories need to be more flexible to accommodate cross-disciplinary learning. Garcia-Garibay added that chemistry community is at a crossroad—it can either circle the wagons around its traditional boundaries or the community can expand to take ownership of newer fields that involve chemistry, including biochemistry and materials science. Trevor Sears from Stony Brook University commented that it is important to work with university administration to explain that the paradigm for teaching science is changing and that classroom and laboratory space needs must reflect that change.

3

Innovations and Barriers

"Assessment and evidence are critical components of innovation adoption."
Thomas Holme

"It doesn't matter whether [students] are going into law, business, economics, or international politics—they have to have a sense of where these technical forces are and how they're shaping the world."
James Anderson

The predominant focus of this workshop was to identify the barriers to improving chemistry education and to highlight innovative approaches to overcome the barriers. Seven speakers presented on a range of topics to enhance the learning experience, including key requirements for education reform and innovative approaches that move away from standard lecture and testing formats; those presentations are summarized in this chapter. Thomas Holme, of Iowa State University and director of the American Chemical Society (ACS) Exams Institute, spoke about some of the challenges in replicating education reform efforts. Clark Landis of the University of Wisconsin, Madison, presented an example of large-classroom reforms and the challenges with reform assessments using traditional metrics. A new approach integrating chemistry and physics curricula into introductory courses was discussed by James G. Anderson of Harvard University. Scott Auerbach of the University of Massachusetts, Amherst, described an approach to undergraduate education that puts context front and center in course design. Michael Cima of Massachusetts Institute of Technology detailed his work developing a massive open online course (MOOC) on solid-state chemistry. Jeffrey Moore talked about the student-centered organic chemistry class he developed at the University of Illinois in Urbana-Champaign. Last, how the design of exams can influence how and what students learn in their chemistry classes was discussed by Angelica Stacy of the University of California, Berkeley.

PROPAGATING MEANINGFUL REFORM IN CHEMISTRY EDUCATION AND THE RELATIVE ROLES OF ENTHUSIASM AND EVIDENCE

What are the barriers that keep successful chemistry education reforms from spreading beyond a local college or university? The fundamental tension, said Thomas Holme, is that teaching is both a personal experience and a corporate enterprise. "It's a personal activity to each of us, but a corporate enterprise because other people care how we do it," explained Holme. He added that anyone who has written a textbook from scratch has experienced this tension. This same tension arises when it comes to educational reforms—everyone is interested in a very broad way in propagating successful new methods for teaching chemistry, but such methods need to appeal to individual teachers. "Enthusiasm can get us in the door," but it must be teamed with an assessment in order for education reform to be effective, argued Holme.

There are resources available to help with education reform efforts. A National Research Council (NRC) report, *How People Learn: Brain, Mind, Experience, and School* (NRC 2000), provides a good foundation and notes the importance to the learning process of prior knowledge, whether that knowledge is correct or not, and metacognition, the ability to reflect on one's own thinking and learning processes. Holme pointed out that a fairly robust albeit young field of chemistry education research was among the disciplines that contributed to a 2012 NRC Discipline-Based Educaton Research (DBER) report from the Board on Science Education. The DBER report also stresses the importance of prior knowledge as an obstacle to teaching chemistry. Chemistry students, for example, have a difficult time envisioning the particulate nature of matter, and this difficulty is often associated with the way it is presented (Cooper et al. 2010). Student conceptions of size scale (e.g., nucleus, atom, molecule, and compound) also present a major challenge in chemistry, and evidence suggests that the ability to understand scale may be the best predictor of success in general chemistry (Gerlach et al. 2011). Unfortunately, chemistry educators are to blame for at least part of this difficulty, said Holme.

The DBER report (NRC 2012) also talks about how to use educational research to impact real-world teaching. The report acknowledges that it is hard to turn basic research into interventions and it is harder still to turn local changes into larger scale change. It also notes that while the National Science Foundation (NSF) has funded professional development activities related to teaching for some time, those activities have largely been self-selecting, leading to a "preaching to the choir" effect (Feuer et al. 2002). The good news, Holme explained, is that the chemistry community as a whole is good at thinking about the diffusion of innovative education reforms, and this is where the relative roles of enthusiasm and evidence come into play.

The classic book on this subject is *Diffusion of Innovation* by Everett Rogers (2003), said Holme, who spent a few minutes summarizing some of the key ideas in the book. The basic definition of diffusion of innovation is "the process in which an innovation is communicated through certain channels over time among the members of a social system." Communication, said Holme, involves convincing somebody who is listening for evidence about key innovation characteristics such as relative advantage, compatibility, complexity, trialability (can a new technology be implemented in steps), and observability. If evidence for these characteristics is not provided, listeners will make up their own evidence and likely conclude that the status quo is better.

Given that an implicit goal is to convince somebody that a new educational innovation is worth adapting, it is important to understand the five stages of innovation adoption. "At the risk of sounding unduly mercantile," said Holme, "we need to understand our targeted customer." The five stages are

1. Knowledge of the innovation
2. Persuasiveness of the innovation—is it better for me?
3. Decision—adopt or reject
4. Implementation—adapt and adopt
5. Confirmation—to keep or not to keep

Enthusiasm, Holme said, plays a key role in the first two stages. Evidence is key in the second and third stages. Assessment is critical for the final two stages. But, Holme cautioned, each of these stages takes place in a social system that is not exchangeable and not particularly amenable. In addition, university faculty will not just be skeptical in the face of enthusiasm, but they will "jump in with their skepticism at every opportunity." It is important to remember, Holme added, that "our colleagues may, or may not, know how to decide about the efficacy of educational innovations they try, but they probably believe they know." This is why assessment and evidence are critical components of innovation adoption. The problem, however, becomes one of getting the data that will serve as evidence in the context of the academic social system.

This is where the ACS Exams Institute can help. The ACS Exams Institute writes nationally standardized exams covering all fields of chemistry and provides resources for outcomes measurement. Recently, the Exams Institute, with funding from the NSF, conducted a national survey of 14,000 professors and instructors in the United States to assess their understanding of assessment terminology and techniques. Holme described the survey and the statistical methods used to analyze the responses. The 1,500-plus survey respondents fell into six clusters of understanding. General familiarity with assessment terms was not high across the six clusters, although analytical chemists tended to score higher than other groups in understanding statistical terms and methods. The lesson here, said Holme, is that everyone needs to be careful when conducting assessments; they need to truly understand what they are doing and what the results are telling them in terms of evidence for whether an innovative chemistry education method works.

Holme noted, too, that sampling in assessment surveys remains a challenge because most studies of educational innovation use convenience samplings of the students who come to the course. Studies also tend to build in bias because disaffected students leave the course before data are collected, an issue of particular concern for those who study MOOCs, for example. He remarked that institutional review boards, which become involved with research on human subjects, place an important constraint on building meaningful control-based experiments. "If we know something is fundamentally better for students, it is unethical to train some of our students with something we know is not good," explained Holme.

Responding to a question from Angelica Stacy of UC Berkeley, Holme acknowledged that it is unlikely to ever have enough evidence to prove an educational innovation is effective, but that it should be possible to have enough evidence to take wise action as to where to go with an innovation. Auerbach then asked what is known about assessing process-oriented, laboratory-type courses, and Holme responded that there are resources available on this subject and added that assessing those courses is more challenging. He added, too, that the ACS Exams Institute has the resources to build assessment tools to help those who are developing educational innovations, but that it will take time to develop tools that the ACS feels are good enough to stamp with its imprimatur.

LARGE-CLASSROOM REFORMS: FIVE BEST TEACHING PRACTICES

The observation that certain groups of students are underperforming in introductory or "gateway" chemistry courses has been an important driver for reforming undergraduate chemistry education at the University of Wisconsin, Madison, explained Clark Landis. Another driver is the desire to increase the fraction of students successfully completing introductory chemistry and who will then take the second semester of the two-course chemistry sequence. Landis and his colleague Ned Sibert also hoped to make teaching introductory chemistry more exciting for themselves. In addition, UW-Madison compiled student data in 2008 and discovered a gap in terms of adverse events, which Landis defined as getting a grade below C or dropping the class, between targeted minorities and the general, nontargeted student population (see Figure 3-1).

Landis described the comprehensive course reform that he and Sibert designed based on what they call the "five best teaching practices" that influence student success in college courses: learning in context, group-based learning, increased time on a task, increased frequency of feedback, and a positive classroom climate (Brower 2009, Cabrera and La Nasa 2005, Treisman and Surles 2001, University of Wisconsin–Madison 2013). These reforms include using concept tests and clicker questions in lectures as ways of making lecture sections more interactive. Online homework, tutorials, videos, and simulations are important elements of the new Chemistry 103 course design, as are peer-led teaching and review. The new course includes spiral curricula, the idea that it is possible to introduce many concepts at a fairly superficial level early on and then return to those concepts regularly, developing them in greater depth each time the concepts are discussed during the semester. Landis explained that spiral curricula work well in conjunction with big, real-world problems.

The primary focus of these changes was to promote active learning in the context of a course that as many as 2,100 students take each fall. The structure of the course includes three lectures, two discussion sections, and 2 hours of laboratory instruction each week. Inquiry-based cooperative learning activities centered on group-oriented challenge problems are used in the discussion sections and in voluntary evening workshops.

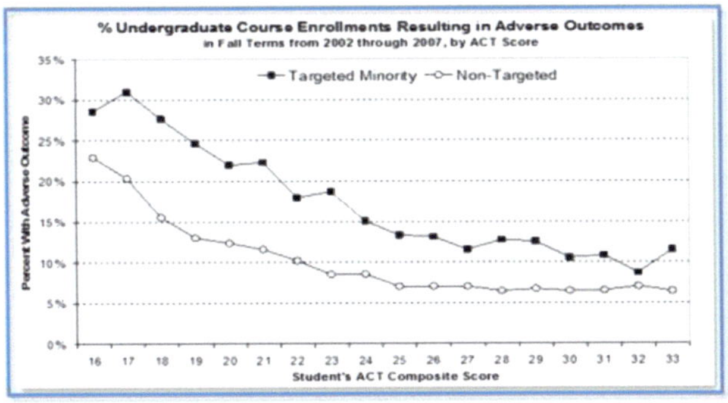

FIGURE 3-1 Gap in rates of adverse outcomes for students across a range of ACT scores. SOURCE: Clark Landis.

From fall semester 2009 through spring semester 2011, Landis and Sibert compared student performance between a reformed and a traditional chemistry course. The study included 189 students in targeted groups and 1,680 students in nontargeted groups in the reformed courses, and 170 targeted students and 1,333 nontargeted students in the traditional sections. Assessment elements included grades and retention; six common questions on the final exam; one common essay question; and student surveys of hours spent outside of class, course perceptions, and student assessment of learning gains. They also monitored gender-based and target group achievement gaps (Seymour et al. 2000).

The results surprised Landis. The reformed course did not appear to affect the achievement gap for targeted students, the fraction of adverse outcomes, or performance on common final exam questions or the essay question. Students in the reform sections did perceive a greater emphasis on collaborative and conceptual learning, worked and discussed more outside of class, attended class more often, and related chemistry more to daily life. They were also more confident in their problem-solving skills.

These surprising results led Landis to ask, "Should we evaluate performance differently?" He cited work from the New Traditions Project, conducted some 15 years ago, showing that students who had been in active-learning sections performed no better than students in traditional sections when tested using standard written exams. However, this study found that "almost uniformly, the students that were in the active learning class were assessed as being better performing in oral exams than the students in the traditional class," explained Landis. "It could be that we just do not have good ways of assessing effectiveness of these methods."

Landis also noted that a study conducted at the University of Colorado showed that the gender gap in a second- or third-year physics class disappeared in the course of writing two 15-minute essays on how a student's values are related to the course. However, when Landis and Sibert conducted the same experiment, they found no differences in performance. He concluded his remarks by saying, "We think value is added by the reformed classes, but we just are not capturing that value in our standardized assessments."

William Tolman commented that he and his colleagues at the University of Minnesota have done similar experiments in organic chemistry class design and also found that performance as measured on written tests did not improve, but that attendance and both student and teacher satisfaction improved. One area that might be showing improvement, he said, is in discovery-based team learning in the organic chemistry laboratory course, though the results are still preliminary. Scott Auerbach thought that one problem with assessing these new teaching methods is that they have different sets of learning goals that the standard assessment techniques are not designed to capture. Landis agreed and noted that there is a good chemistry concept inventory being developed that he is eager to use as an assessment tool.

Anne McCoy wondered if differences might start showing up in later courses, where students who had taken the reformed classes might show better retention of the concepts they learned in the new chemistry sections. Landis said that he would like to run such longitudinal studies, but funding is an issue. He reiterated the need for longer-term support when Jodi Wesemann asked what was needed to keep these reform efforts going.

Jeffrey Reimer from the University of California, Berkeley, asked if anyone had conducted studies comparing sections taught by white male faculty and those taught by targeted minority faculty. Landis said he did not know of any work in that area but noted that he and his colleagues found no difference in performance between students taught by male versus female faculty.

TEACHING INTRODUCTORY CHEMISTRY WITH A MOLECULAR AND GLOBAL PERSPECTIVE

James Anderson described Harvard University's efforts to infuse chemistry and physics into an introductory chemistry course. Their goal is not to recruit more chemistry majors, said Anderson, but to develop a chemistry curriculum that is evolving to keep more students engaged in the physical sciences as a whole and to make the rest of the student population more aware of and appreciative of the physical sciences. The reason for that emphasis, he explained, is that the physical sciences are playing a central and critical role in solving the major problems facing human civilization today.

Harvard's Physical Sciences 11, Foundations and Frontiers of Modern Chemistry: A Molecular and Global Perspective, reflects the idea that introductory courses in both chemistry and physics have been taught separately and without a compelling context. The Physical Sciences 11 course is based on the premise that decisions on what university graduates face in their academic career directly relate to what they take in their freshmen year. "If the separation between science and society occurs in the freshman year, it's irreversible for that generation," said Anderson.

Chemistry and physics faculty are both to blame for the lack of appreciation for and understanding of the physical sciences because the courses they teach create an exclusive club of students who can excel at these subjects instead of an inclusive group of students who understand the basic concepts of the physical sciences. In contrast, said Anderson, the life sciences have clearly demonstrated how important they are in the larger context of society. He explained that by context he meant linking the essential concepts of chemistry and physics to their connection with the big problems that intrigue students today—energy, human health, national security, climate change, and others. He noted that while there are significant differences among universities—yes, he acknowledged, the students at Harvard, MIT, and Caltech are somewhat different—there is a general pattern common to all: attrition from the sciences during and following the freshman year (see Figure 3-2).

For the first few decades after World War II, only 10 percent of entering undergraduates completed their baccalaureate degrees in science, which met the demands of graduate and medical schools. Today, however, "this zone of scientific and technical illiteracy has now become a fundamental problem that we have to deal with because of these issues of national security and competitive economic considerations on a global scale," said Anderson. The problem of driving scientific and technical literacy to a level where 95 percent of graduates are scientifically and technically literate is what prompted Harvard to completely revamp freshman chemistry, he said.

The current strategy in introductory chemistry, said Anderson, is to present lectures and text material that covers the basic formalism and theory, followed by problem sets and exams. Solid evidence shows, however, that there are two basic failures with this "formalism first" approach to teaching. First, he said, it results in "disembodied knowledge"—students cannot attach the knowledge to a context or their past experience, and so it is largely meaningless symbols and

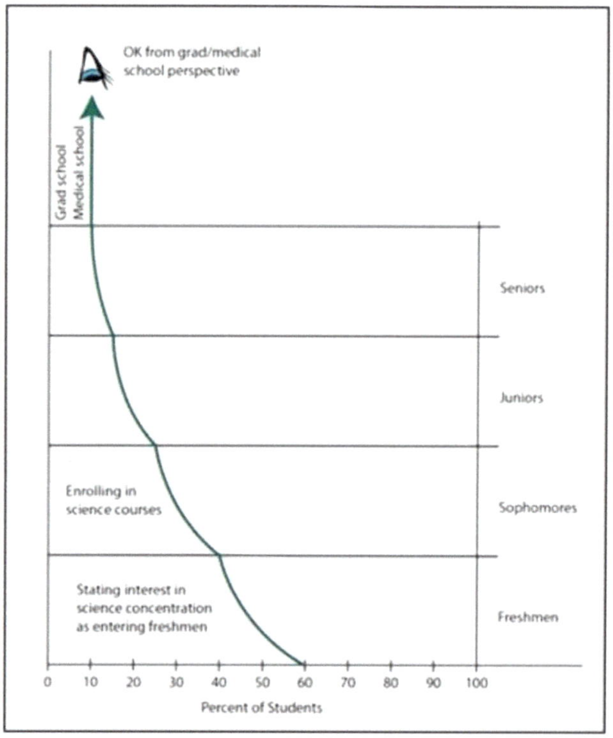

FIGURE 3-2 Attrition of undergraduate students from science majors. SOURCE: James Anderson.

facts to memorize. Second, as Nobel laureate and noted science educator Carl Wieman has shown, knowledge obtained this way is filed away in the brain in a separate compartment and building links to that compartment after the fact is much harder and less effective than if it had been filed correctly from the start.

Anderson and his collaborator, Harvard physics professor Efthimios Kaxiras, have taken an approach to addressing these problems with a strategy that links concepts with context (see Figure 3-3). As an example, electrochemistry is a great way to teach about Gibbs free energy, electron flow, electromagnetism, and chemical transformation when it is connected to a context of the electric car. In writing the textbook for this course, Anderson and Kaxiras used case studies that are broken into their parts, analyzed, and reassembled. When Anderson gave the first lecture of Physical Sciences 11, 25 students were present. By the fourth lecture, there were 125 students in the lecture hall. A year later, around 300 students had completed the class, which is notable because Physics 11 is a considerably more difficult course than the alternative class the students can take, Anderson said.

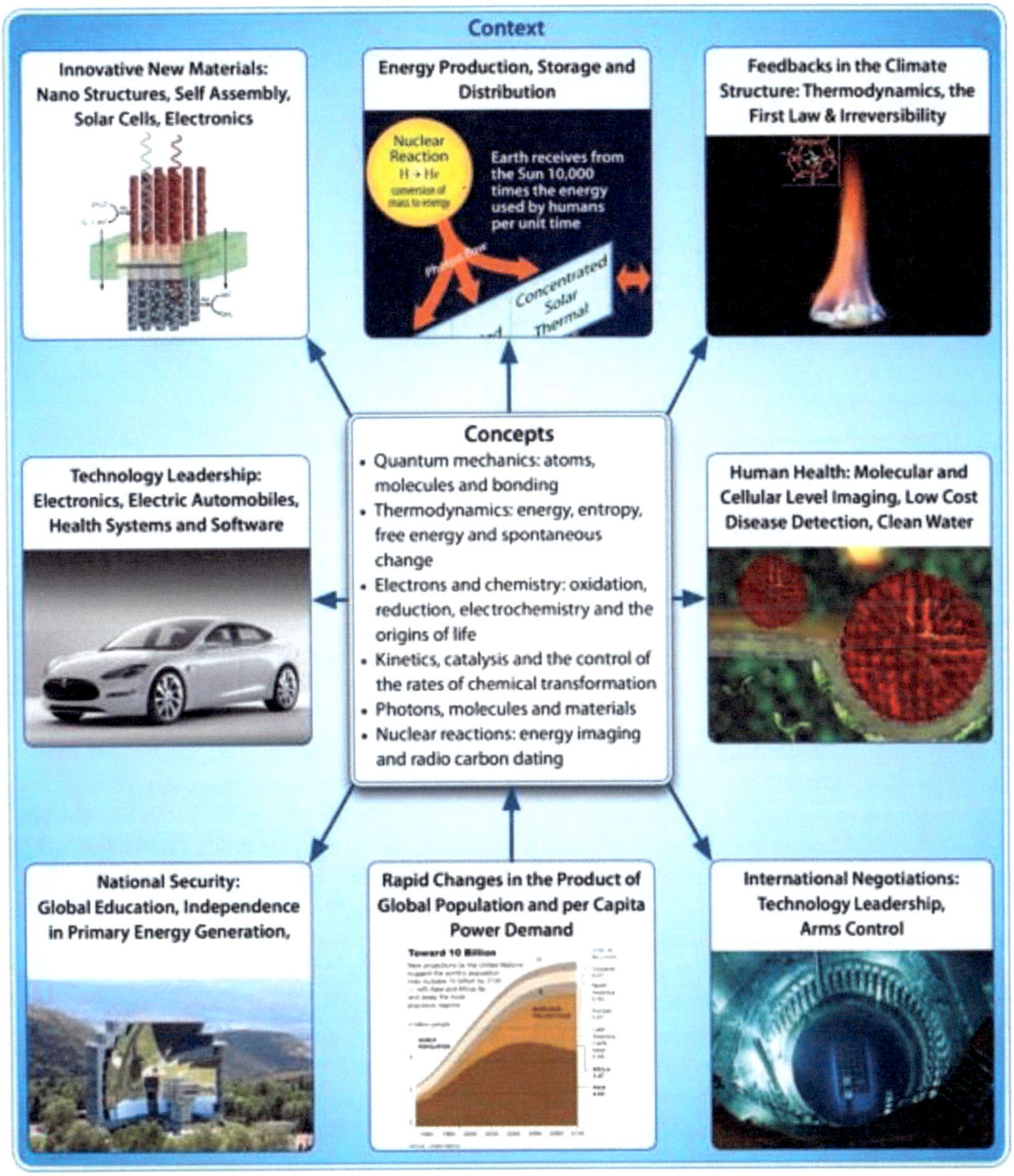

FIGURE 3-3 Linking concepts to context is the guiding principle for a new introductory physical science course taught at Harvard University. SOURCE: James Anderson.

David Harwell asked if faculty at other institutions are likely to adopt this approach to teaching, given how much more time and effort it takes. Anderson replied that adopting a concepts-with-context approach for classes at other institutions will be a challenge. This is one reason that he and his colleagues are putting so much effort into developing the textbook and associated materials, which will be made available to students outside of Harvard for $15. Anderson asserted that even though this style of teaching is more difficult, "I cannot imagine going back and teaching it the old way." Cardillo noted that faculty at other institutions have tried this course with the textbook that the Harvard team authored and he characterized the results as extraordinary. "It's starting to catch on," said Cardillo.

TODAY'S STUDENTS AND TOMORROW'S LEADERS: INTEGRATED CONCENTRATION IN SCIENCE

The motivating factor behind Scott Auerbach's involvement in science education reform is that he does not believe that the nation is training its students to succeed in the important areas of science that are crucial to the future of the society. He and a group of colleagues from several academic departments at the University of Massachusetts have responded to this shortcoming by developing the Integrated Concentration in Science (iCons)[1] program, where groups of students with diverse backgrounds work in teams to develop solutions to today's major problems. The program's mission, explained Auerbach, is "to produce the next generation of leaders in science and technology who have the attitudes, knowledge, and skills needed to solve the inherently multifaceted problems facing the world."

In developing the iCons program, faculty developed a long list of desired student outcomes and organized them into a set of 10 learning goals. For example, the first goal is that students will be able to critically evaluate societal challenges and possible scientific solutions, and another goal is to develop quantitative understanding of societal problems and solutions. In general, the students do not have the ability TO discern the quantitative regime, and so an important aspect of iCons for faculty is to provide those skills. A third learning goal is for students to be able to design, carry out, and interpret valid scientific studies related to societal challenges. Connecting the dots to societal challenges—that is the key, said Auerbach.

After developing the learning goals, the iCons faculty realized that the skills that students develop as a result of achieving the learning goals are applicable outside of the classroom. That is, the skills may be interchangeable with the key cognitive, intrapersonal, and interpersonal skills that the NRC noted as being critical to success in life and work in the 21st century (NRC 2012). Auerbach and his colleagues were also gratified to see that their learning goals mapped onto the crucial elements of successful science, technology, engineering, and mathematics (STEM) programs identified by the President's Council of Advisors on Science and Technology.

In practice, iCons is an 18-credit, 4-year concentration that does not replace the student's major (see Figure 3-4). "Every student in the program is a major in some field of engineering or science," explained Auerbach. "That major is the cake; iCons is the icing on that cake, and that icing is in the form of case studies, lab work, and research." Every iCons project involves a case study (NRC 2011b) and has four essential elements: problem-based science and engineering (Gijbels et al. 2005), multidisciplinary student teams, student-driven collaborative learning, and reflection and self-assessment. As an example, he discussed a case study on high fructose corn syrup that was used to teach carbohydrate chemistry. The case study started with two articles in the popular press, one in 2010 that reported on a study showing that high fructose corn syrup promotes weight gain, the other 2 years later purporting that high fructose corn syrup is no worse than table sugar. The students were charged with getting to the bottom of this conflict, which involved learning not only about carbohydrates and carbohydrate metabolism, but also about the limitations of studies and how to design a new study that addresses those limitations. At the end of this case study, the students reflected on what they had learned, how they had learned it, and what they would do next in terms of gaining more knowledge on the subject and putting their ideas into practice.

As freshmen, iCons students learn about teamwork and take on numerous case studies. Prior to starting their second year, students choose a theme for their future work, either energy or biomedicine. As sophomores, they take theme-specific courses focusing on communication, reading, writing, speaking, and debating on the issues that are relevant to their chosen theme. In year three, students move into the laboratory and begin designing experiments using cutting-edge equipment (when relevant) to address real-world problems as part of a research group. As seniors, students will engage in an interdisciplinary research project, complete a portfolio, and write their honors thesis. iCons is currently in the third year of the program and the three cohorts include 130 students from 20 different majors from the colleges of engineering, science, and public health.

To determine whether iCons is meeting the learning goals, faculty have developed eight assessment instruments that pair three categories of assessment: formative/summative, qualitative/quantitative, and generic/targeted. So far, the program has implemented six of these instruments to assess iCons. For example, one weakness of iCons was found in the Student Response to Instruction Instrument (SRTI), which is a summative, quantitative, generic assessment tool. The SRTI showed that students were receiving insufficient feedback

[1] See http://www.cns.umass.edu/icons-program/.

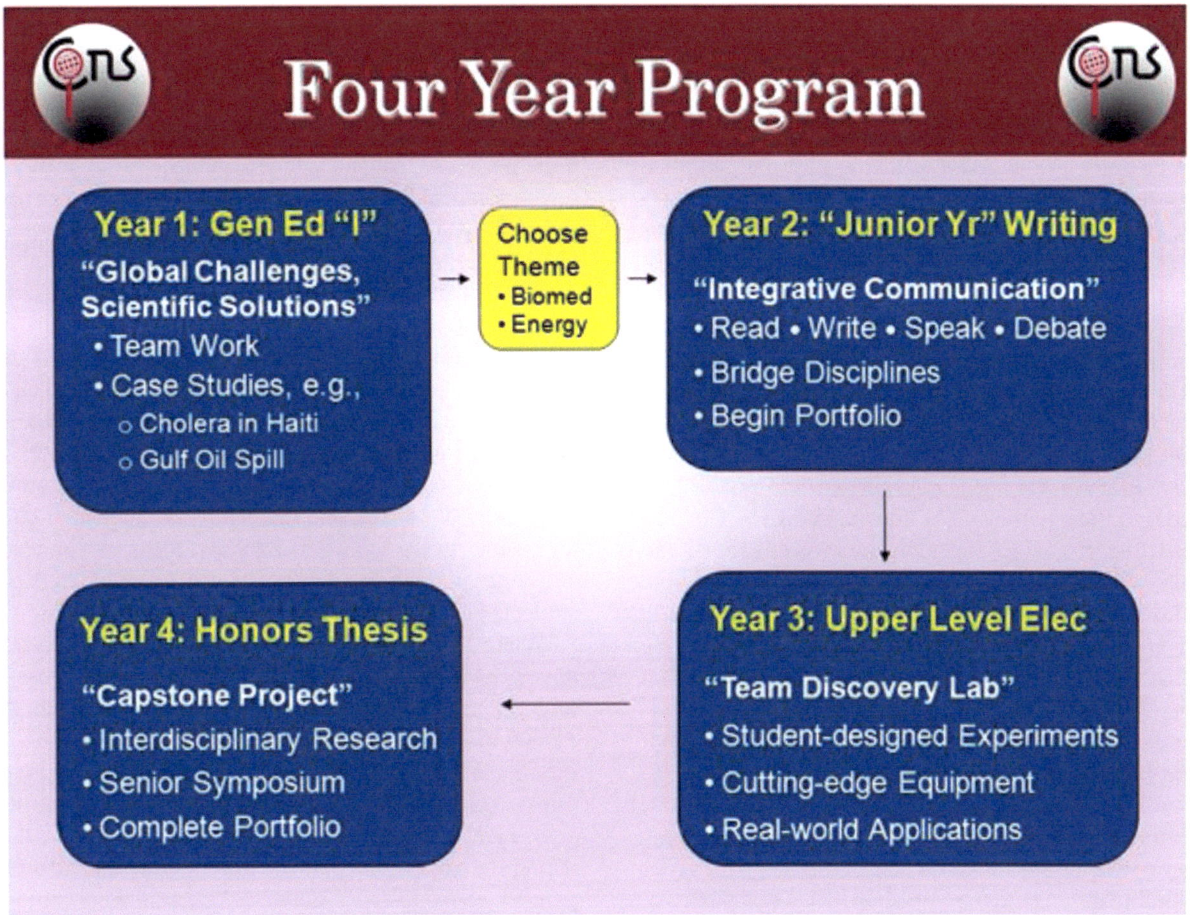

FIGURE 3-4 The 4-year iCons program at the University of Massachusetts. SOURCE: Scott Auerbach.

on their performance in the class, and so the program has included more opportunities for feedback. Auerbach noted that the response signature of the assessments has been repeatable across the first two cohorts for the first iCons course. In terms of broader impact, there are now student-driven team projects in general chemistry, organic chemistry, physical chemistry, and integrative graduate education. Faculty who have participated in iCons also report that they have changed the way they teach their other courses. The take-home message is that education is not about "filling a bucket," but about "lighting a fire" through context, Auerbach said. "Tell me and I forget. Teach me and I remember. Involve me and I learn."

Jay Labov from the National Academy of Sciences asked if iCons is changing the culture at the University of Massachusetts in terms of promotion, tenure, and teaching. Auerbach replied that the program has not changed how departments view the metrics for tenure, and as a result, he would never ask a junior faculty member to teach in iCons until the minute after they received tenure.

After declaring his support for this type of program, John Kozarich of AxtiveX Bioscience asked about the rigor of the education that the iCons students receive. Auerbach replied that the goal of iCons is to instill passion in the students for a given idea that then prompts the students to drill down deeply into a subject. He also acknowledged that the program does face the challenge of balancing depth of learning with depth of exposure.

When asked how students fit these courses into their busy schedules, Auerbach said that the iCons courses are designed to substitute for existing courses. This is expensive, however, because the original courses still need to be taught for students that are not in iCons. "The only way this will work in the long term is if companies love the product that we have so much they are willing to give us money to continue training students in this way." He noted, too, that there is a 20 percent dropout rate from the program because of scheduling conflicts. "Finding time to get students from 20 majors together at one time is a big barrier," he said.

A participant asked if iCons is considering how to turn some of these student ideas into real-world activities. Auerbach said that the plan is to work with the business school to develop a venture capital track over the next couple of years.

ONLINE EDUCATION AND MOOCS

Working from the underlying theme that the chemical bond determines properties, and with an emphasis on linking basic concepts with applications, MIT's unique first-year chemistry course teaches students fundamental chemical principles through the solid state. The hypothesis around which the course is based is that students will find it is easier to learn, understand, and, most importantly, use chemical principles if they can relate them to the solids around them. Michael Cima has now adapted this course, which has been offered as an option for meeting MIT's general instructional chemistry requirements since the 1970s, to serve as a MOOC on the edX[2] online platform. The course, known as 3.091x,[3] is offered free of charge; includes homework, exams, and a final exam; contains the same intellectual content as the classroom-based MIT course on which it is based; and is a certificate-earning rather than credit-earning course. He added that the course is taken by a cohort of students that work together.

Cima explained that 3.091x is an engineering course, not a chemistry course, and that affects the way the students are tested. "We are assessing students not on what they know, but what they can do, and they do a lot of calculations, which turns out to be an advantage for an online course." After demonstrating the class to the workshop, Cima noted that the experience of doing a screencast was much more interesting for him as a teacher than standing in front of a large lecture hall. He noted that classes on the edX platform are nothing like traditional online courses in that there is no 50-minute lecture. At most, each lesson consists of a 10-minute lecture segment on a single concept followed by self-assessment tools that are graded immediately and serve as a reality check for the student.

For MIT, the reason to offer an online version of an established course was simple: research consistently shows that learning outcomes are about the same for a residence-based course and an online version, but that when the versions are combined, the students do better on both courses. The goal, shared between MIT and Harvard, is to use both versions simultaneously, and Cima was going to do just that starting with the fall 2013 semester.

Course development took considerable resources in time and money. Cima explained that course preparations began in June 2012, 4 months prior to 3.091x's launch on October 15, 2012. Producing the course took one teaching assistant working over the summer, one full-time and one half-time edX person, production and engineering staff time that totaled 2.5 full-time equivalents, a part-time administrative support person, five paid forum moderators to answer questions posted by students, four volunteer community teaching assistants, and two to five beta checkers. The lecture video derived mostly from his 2011 lecture class. Some 280 lecture segments, 65 screencasts, and six additional video segments were incorporated into the final version of 3.091x. "It was a huge amount of work, and I spent the bulk of the summer getting ready for this," he said. Now the course is offered, he goes online himself most mornings and answers student questions. He noted, though, that the unpaid community teaching assistants, who are people taking the class in locations worldwide, have been an amazing resource. In fact, Cima has asked one of them—a high school chemistry teacher in the United Kingdom—to serve as a teaching assistant for this coming fall semester. Six of her high school students have also taken the class, he added.

When the course was offered, nearly 29,000 people registered for it. Over 3,400 took the first test, almost 2,200 took the second test, and 2,148 took the final exam. About 15,000 of the students were using the materials throughout the course and, on the basis of results of an exit survey, Cima thinks that the bulk of these people are taking a chemistry course and using 3.091x as a teaching supplement. Of those taking the course, 13 percent were graduate students, 28.9 percent were university students, 1 percent were community college students, and 9.7 percent were high school students, some of whom want to know what taking an MIT class entails. The biggest surprise, he said, was the large number of teachers—almost 9 percent—who took the course, and he has corresponded with many of them. Some 3 percent were K-12 teachers and over 5 percent were university or community college teachers. It may be feasible, he said, to offer these kinds of classes for professional development credits for high school science teachers.

In a retrospective look at outcomes, which he did by designing the residence-based final exam in such a way that he could do a select group of measurements with the online final, it appears that the online students outperformed the residence-based students. Cima believes this "troubling" finding resulted because it may be possible to do a better assessment of student learning in an online setting than in a classroom under time constraints. One of his goals for the future is to develop improved assessment tools.

Another task Cima faces is determining how best to maximize the benefits of having all of the developed content and assessment tools, particularly as he works to integrate the online and residence-based classes. He noted that the decision to integrate the two is highly political at MIT and he spent considerable effort building support and getting approval for this change. This coming fall, Cima is going

[2] This nonprofit organization offers MOOCs and interactive online classes for a variety of subjects, including some STEM subjects. It was founded by individuals from Harvard University and MIT.

[3] See https://www.edx.org/course/mit/3-091x/introduction-solid-state/591.

to conduct an experiment that will consist of replacing the course texts with the online content and developing lectures to take advantage of the new "text." The course will be structured around two 1-hour recitations per week, and assessment will consist of 37 proctored online quizzes that the student will complete within a specified time window. Each quiz will represent a learning outcome measure, and if a quiz is not answered correctly, the student may take the quiz again and for as many times as they want within the time window. There will be a 24-hour lockout between quiz attempts. Each learning outcome quiz will be selected randomly from a group of many related problems. Cima noted that he used this quiz format with the last midterm of the online class without any problem.

A participant asked what the implications of MOOCs might be for university education, and Cima said that while he has no real idea, it could be that the freshman year is spent off campus taking their required foundational courses online. He was then asked if the videos will still be fresh in 5 years or if the course will have to be continually redesigned at significant expense. Again, Cima replied that he did not know what the future held but noted that it would be easy and relatively inexpensive to replace videos with screencasts. He also said replacing content is simple once the course is constructed, claiming that he can use a new software tool to replace content from his desktop computer in 5 minutes.

A participant asked if MIT was considering faculty diversity in its plans to create a catalog of MOOCs. Cima said that the university takes this challenge very seriously since the MOOCs do represent the face of the university and that face has changed considerably over the past 30 years. He noted that the introductory mechanical engineering MOOC is taught by two female faculty.

Wesemann asked if the positive experience he had creating 3.091x was having an effect on other faculty. Yes, said Cima, and in fact, each engineering department is hiring a person who will be dedicated to helping faculty convert their courses to an online format for use by residence-based students. In response to a question from Cardillo about other companies offering online courses, Cima said that his impression, as well as that of other faculty he has talked to, is that there is a wide diversity in terms of the quality of these courses. What he likes about the edX format is that it is based on an open-source system and developers are taking advantage of that to develop assessment tools for community use.

DEALING WITH RISK, FAILURE, AND UNCERTAINTY

The challenge that Jeffrey Moore is tackling at the University of Illinois with his instructional experimentation is to center instruction on the individual learner in what he called the "kilostudent" organic chemistry classroom consisting of a diverse set of nonchemistry majors who, in most cases, are taking their last formal course in chemistry. His approach to meeting that challenge has been to reconfigure the learning outcomes of the course to match the grand challenges in science education enumerated in the 2013 special issue of the journal *Science* (McNutt 2013) and to design a curriculum that uses theory webcasts to present concepts, pressure-point problems to immerse the student in experience, problem-of-the-day discussions, and peer-to-peer tutoring.

One of the key features of the course is its use of online electronic homework, which for organic chemistry works out wonderfully with machine-read computer drawings that can be automatically graded using a programmed graphical language. There is no textbook in the course, though there is a set of course notes that accompany the 5-minute webcasts that the students view before attending a discussion session. The discussion sessions are held in a computer lab that can accommodate 55 students at a time, or via an Internet connection at two times daily. At the end of the discussion session, the students are presented with a pressure-point problem that they have 5 minutes to solve. Successfully answering the question yields bonus points that are applied to the next exam.

These complex problems, explained Moore, are designed to take students into uncharted territory and force them to take risks and fail, just as scientists do in the real world (see Figure 3-5). He called this "taking off the training wheels," and said it teaches students about failure and how to respond to that failure. The problems are nonalgorithmic and multifaceted, and multiple steps are involved in solving them. Students are forced to use creative processes to generate a variety of initial-guess solutions and to develop a strategy to initiate a solution. They are also allowed to access the literature or any other online resource—except communicating with another person—to solve the problem.

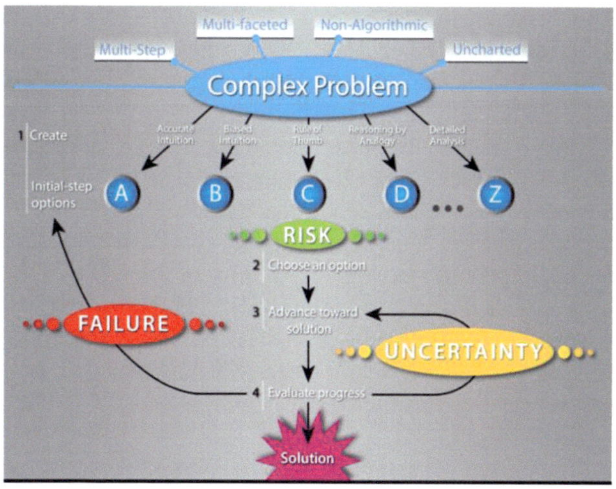

FIGURE 3-5 Learning through experiences that mimic the real world. SOURCE: Jeffrey Moore.

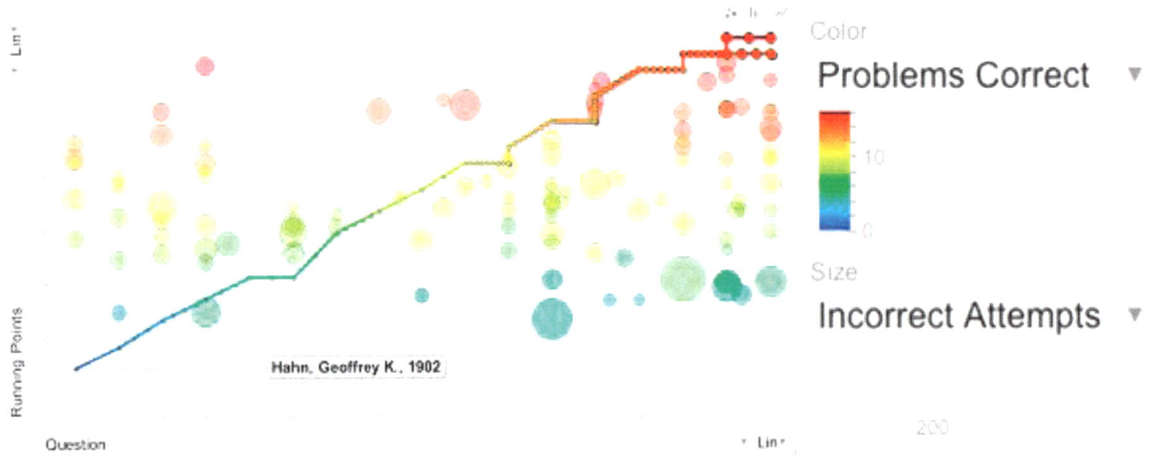

FIGURE 3-6 Graphical tracking of exam progress. SOURCE: Jeffrey Moore.

No points are given for an incorrect answer, but students are allowed to continue trying to solve the problem until they get the correct answer. Because the students are developing and refining their solutions online, the computer can track the progress they are making toward the solution. When plotted, these data provide a picture of how the exam is going and how each student is doing in real time (see Figure 3-6).

One of the main components of Moore's course is peer-to-peer tutoring aided by the Internet. "We realized through some of our assessments that one of the most important ways that students were learning was not by using the videos, not even by the problems that they were doing, but by the interactions that they were having with other students," said Moore. The way this worked was when a student answered a problem correctly, they would be placed in the tutoring pool for that problem. Students who answered a problem incorrectly and who wanted help would then go to the tutoring pool, where they could select a peer tutor to help them. At the end of the process, both students then recorded video reflections of the outcome of that tutoring experience.

"We do not monitor every single one of those videos, but we do keep track of the clusters of people and the information flow, the diffusion of information," he explained. Using this information, Moore can create a network map of tutor–tutee interactions (see Figure 3-7). One of the surprises that came out of the analysis of the video reflections was how much value the students placed on peer-to-peer tutoring from both the tutor and the tutee. Watching video lectures online was deemed the least valuable method of learning by the students.

Moore has also implemented a semester-long group project in which four-student teams select a small organic molecule, typically a bioorganic molecule, and use the literature to propose a mechanism for how this molecule might be made, its properties, and other relevant information. The idea behind this project is to promote the development of professional scientific skills, and it appears to produce gains in literature searching, scientific writing, and critical reading, though there is evidence that sustaining these gains requires what he called a "super teaching assistant" or an intensely devoted instructor.

In summary, Moore said that this revamping has successfully flipped the two-semester organic sequence without significantly increasing the load of the teaching assistants. "I can say from the data we've collected that we've done no harm," he said. "We have not improved things in terms of

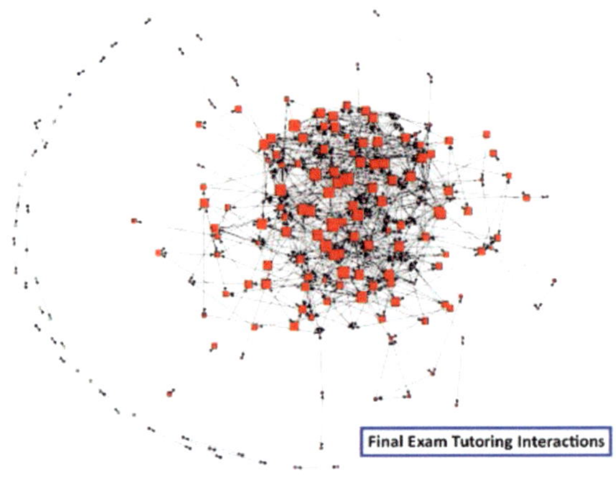

FIGURE 3-7 Tutor–tutee interaction network. Arrows illustrate the hypothetical transfer of knowledge from tutor to tutee, and the nodes are scaled by how central they are to the network. SOURCE: Jeffrey Moore.

these learning outcomes and objectives, but at the same time, learning is robust." In his view, machine-graded homework, discussion problems, daily pressure-point problems, and exam problems offer students many more opportunities for practice with feedback, and that testing with real-time feedback might measure students' ability to diagnose their errors in a pressure situation. He reiterated that peer-to-peer tutoring appears to be a valuable tool for student learning and added that it requires little effort to implement.

Jeffrey Reimer asked if the students are aware that whatever they do for this course is analyzed and therefore not private. Moore replied that he has received institutional review board approval and that the students agree to participate in the data collection and analysis process. Session chair Emilio Bunel asked if other faculty were involved in teaching this course. Moore said he did have one colleague involved who has since moved to another university, but his impression is that faculty are intimidated by the amount of work that went into developing the course. There is an instructor for the first-semester organic chemistry class who has completely embraced this approach and has abandoned the traditional lecture approach.

Luis Martinez from Rollins College asked about the importance of training for the teaching assistants and if their experience teaching in this format had any impact on their development as future teachers. Moore said that he is careful about selecting teaching assistants for this course who feel comfortable with technology. He has not yet assessed what the teaching assistants are getting from this experience but he thought that would be a good idea to look at going forward. What he has heard from the students who take the class is that they value how much they learn from acting as teachers in the peer-to-peer tutoring process and how much they now value teaching.

WHAT GETS MEASURED IS WHAT GETS LEARNED: ASSESSING STUDENT UNDERSTANDING

One of the problems in the way students are taught science today lies with the way student learning is assessed, said Angelica Stacy. Legacy chemistry exams, she stated, are designed to select for those students who are good at memorizing. "Teachers are a good delivery system if all you want students to do is memorize," she said. "Teachers ask questions on exams that students can answer, and everyone feels good." The result? "If you don't assess what is important, what is assessed becomes important." Stacy quoted Sir Ken Robinson, an internationally recognized author and leader in education, who said that the way most classes are taught today is turning education into "hours of low-grade clerical work."

Stacy is optimistic that this sorry state can change because of the new climate of possibilities that starts with the Next Generation Science Standards (NGSS)[4] that talk about science practices, crosscutting content, and core ideas. Evidence of coming changes include the new Advanced Placement tests that the College Board is releasing and the redesigned MCAT^{2015} exam. She echoed the sentiment shared by other individuals at the workshop that teaching methods should be revamped based on the growing body of research that shows how students learn.

At UC Berkeley, Stacy and her colleagues have started writing introductory chemistry course exams differently, and they conducted a study to determine the effect of these changes. As part of this study, they asked students at three times during the semester what they were doing outside of class to prepare for the four exams given during the semester to better understand how they were using resources when they were directing themselves and how they changed their study strategies as they realized that the tests were less about memorization and more about understanding and applying concepts.

The students fell into four groups. She calls the lowest-performing group the fact gatherers—the students who memorize independent facts and then are confused by their low grades given the amount of time they put into their studies. The next group, which performed slightly better, learned procedures—they absorbed information and made small connections, but still relied on others or course materials for answers. Students in the third group work at confirming their understanding. They evaluate information and question why, work more independently as learners, and try to give explanations. These students are trying to understand if they are thinking about a problem correctly. The fourth group, which Stacy characterized as amazing, thinks about chemistry all the time. These students are applying ideas, taking in information and questioning why it is true and how it helps to explain the world around them.

Unfortunately, said Stacy, most of the bright, motivated, talented students who come to the University of California, Berkeley, fall into the first two categories. Over the course of the semester, the numbers do shift, with more students moving into group three and a few rising into group four. The big shifts, she noted, do not happen until after the second exam when the students realize the first exam was not a fluke. Students who improved became more active learners and they moved away from just reading the course textbook and started working with peers and asking more questions. Students who did not improve remained passive learners who stay focused on reading text and were not reflective when studying. The students reported that they made changes to their study habits because the exam questions made them apply what they were learning—they could not memorize

[4] The NGSS are new K-12 science standards to provide students an internationally benchmarked science education. The NGSS are based on the NRC report *Framework for K-12 Science Education* (NRC 2011a).

an algorithm to solve the questions and expect to do well on the exam.

Stacy acknowledged that exam design is challenging when teaching large lecture classes. For example, she has 1,300 students to teach and a limited number of graders to help her. "The trick is, can we start to learn how to write better multiple-choice questions," she said. One approach is to use data and observations as the basis for test questions and then structure the classroom experience to develop the skills to explain data, to find patterns, and to understand how to control variables when comparing different pieces of data. She and her colleagues have designed their chemistry curriculum around core ideas in chemistry that are similar to those in the Next Generation of Science Standards: matter, change, energy, and light. She described several examples, including one that uses smells to explore molecular structure and properties. Stacy presented her students with a table with properties of four chemicals found in spices with strong smells or tastes: vanillin, eugenol, zingerone, and capsaicin. The properties given were flavor, molecular formula, structural formula, melting point, boiling point, and water solubility. Students were also provided with space-filling models of the molecules. Along with this information, to the students, they were asked practical questions like "why in the world can I smell vanillin so well?" and "I can't smell capsaicin, but when it gets into my mouth it doesn't go away. Why is that?" The exam questions are designed to combine data and observations in such a way that the students must use core concepts in chemistry, like intermolecular attractions, to explain the data.

In closing, Stacy argued that the field needs to do research-based redesign of undergraduate chemistry courses. "I think it begins with the assessments," she said. "They have to promote understanding as opposed to memorization or we're really not getting to where we want to be. We have to use students' ideas and experiences to build knowledge. Let them observe. Let them explore the data. You'll be amazed at what ideas they do come up with."

YuYe Tong from Georgetown University remarked that this work shows the value and challenges of moving students from being passive to active learners and of emphasizing concepts over content. Sarah Green from the Michigan Technology University commented that it will be interesting to see what the impact will be of the NGSS, which stress concepts and problem solving as opposed to rote memorization, as those students enter the undergraduate chemistry curriculum. Stacy noted that one effect of using these new course designs to teach college students is that they will become the next generation of teachers and professors, creating what could be a virtuous cycle in science education.

In response to a question from Cardillo about the extent to which these innovative courses have been adopted by other faculty, Stacy said that she has been joined by two of her colleagues and they now coteach this course, where they trade off lectures and critique each other. Other faculty have seen how enjoyable it is to teach chemistry in this manner and are working to change their course design too, noted Stacy. Anderson added that he has strong support from his department for continuing to revamp the introductory physical sciences class, but that other faculty are still in wait-and-see mode because of the amount of work that he and Kaxiras have had to put into both redesigning the course and developing the accompanying text and other course materials.

4

Industry Perspectives

"We need scientists that understand how to think, have the technical background, are inquisitive, but who are not necessarily so focused on a particular discipline."
Shannon Bullard

"There are more graduates who are specializing very early, which limits their potential to learn and grow within an industrial organization."
Francine Palmer

A significant number of undergraduates that complete a baccalaureate in chemistry do not go to graduate school or medical school but enter the chemical or pharmaceutical industry. In considering the need for changing the way students receive chemistry education, it is important to consider industry's perspective and ask if major employers of these students see a need for change. To address that question, the workshop participants heard brief presentations by four people with an industry perspective. An open discussion followed remarks by Shannon Bullard, program manager in the human resources department of the DuPont Chemical Company; Francine Palmer, research and innovation director for Solvay Corporation; David Harwell, assistant director of career management and development at the American Chemical Society (ACS); and Robert Peoples, executive director of Carpet America Recovery Effort.

PREPARING STUDENTS FOR THE INDUSTRIAL LABORATORY

From her perspective of leading DuPont's recruiting program for bachelor's degree scientists and engineers, Shannon Bullard does not believe there is an urgent need for dramatic change in the undergraduate curriculum. She does, however, see some opportunities to make improvements, particularly in terms of providing graduates with technical flexibility. Today in industry, customer demands change and as a result DuPont needs its associate investigators, as its bachelor's degree researchers are called, to have the intellectual confidence and skills to move smoothly between different areas of the company. "We need scientists that understand how to think and have the technical background and inquisitiveness, but who are not necessarily so focused on a particular discipline," she explained. The ability to think and problem solve are key as the company looks at how it can contribute to solving those bigger problems in the world that were mentioned in the previous session.

Another area that deserves more emphasis, said Bullard, is that of internships and undergraduate research. Having laboratory experience and putting into practice what they learn in the classroom give students a big advantage when they come into industry.

Francine Palmer said that in her view, learning the fundamentals of chemistry is still key. "We see in our hiring that there are more graduates who are specializing much earlier, which limits their potential to learn and grow within an industrial organization," she said. It is not bad that graduates are coming out with strong skills in the biologically oriented chemical sciences or material sciences, but that they still need that broad understanding of chemistry fundamentals.

Also important, she said, are the so-called soft skills—collaboration and communication—that students can learn in class but more often learn through research experiences, internships, and co-op-type programs. "We encounter many really clever students that are unable to get their opinion across or formulate responses, which makes it really hard in a large research community to be able to collaborate," she explained.

Three particular groups concern David Harwell in his role as director of career programs at the ACS: students, displaced workers, and long-term unemployed workers. Students are at the top of his list because their unemployment rates across the field and all degree levels stand at 13.3 percent. For chemists with only a bachelor's degree, the unemployment rate is 14.6 percent a year after graduation. In contrast, the unemployment rate for displaced chemists is just over 4 percent. The difference between these two groups is experience, said Harwell. The field needs to create more opportunities for internships and other avenues for students to gain practical experience. Undergraduate research is a good alternative, he added.

One area that industry stresses that does not receive much attention in academia is safety and the culture of safety, an idea that Palmer agreed with strongly. Another is working in teams whose composition is always in flux as a project moves through various stages of development. A scientist with a given skill set may be reassigned many times over the course of a career to new teams that need that skill set, and he reiterated Bullard's comment on the need for technical flexibility. Recent graduates also lack networks, or at least they think they do, said Harwell, and chemistry needs to start teaching its students how to tap into the network of former students and former undergraduate research group members. Harwell agreed with Palmer's call to give students more exposure to the soft skills they need to succeed in industry.

Robert Peoples reiterated Harwell's comment that industry is concerned about the lack of safety training for undergraduate students. Many companies have accepted the fact that they need to invest time and energy into developing safety training courses for their new bachelor's degree employees. One addition that he has long believed is needed in the undergraduate chemistry curriculum is a seminar program that brings in industrial scientists to speak with students. He also said that chemistry curricula should include an emphasis on sustainability because industry is acutely tuned into this as a major aspect of competition and future growth. He also stressed the need to develop better communication skills in chemistry students and to establish better mentorship programs.

DISCUSSION

Scott Auerbach, from the University of Massachusetts in Amherst, supported the importance of expanding internship opportunities for undergraduates, but noted that "if we rely on them to teach students the skills that they need to succeed, that's a cop-out. I think we need to be thinking about how we can create different kinds of educational opportunities on campus that are as close as we can to internships." That need, he said, creates a conundrum where credit hours need to be devoted to working in teams, working on bigger problems, and practicing the art of communicating not just to scientists but to nonscientists as well. At the same time, as Palmer said, students need to master the fundamentals of chemistry. "The decision that we have to make is to determine the critical mass of time that we need to be spending teaching these other skills so that we can build them into the chemistry curriculum," he said. Bullard agreed with this assessment but not that it was an either-or solution. She believes that there must be approaches to teaching both, and perhaps those might be found by looking to applied fields, such as food science, that have had years of experience developing curricula with that balance.

Kozarich and Shulman both thought that students need to have some exposure to interviewing skills, which is in a sense an extension of problem solving. Shulman thought that these kinds of "employability skills" could be incorporated into the new requirement in the ACS guidelines that call for students to have a capstone experience. Shulman also asked the panel if the salary premium that chemical engineering graduates receive compared with chemistry graduates is a result of the former having more of these employability skills. Both Palmer and Bullard agreed with that statement completely. Palmer noted that the chemical engineering graduates she hires have much more experience in collaborative problem solving and in presentation skills because those are emphasized in the chemical engineering curriculum. Bullard added that the training focus in chemistry is on independent research in a specific area, not interdisciplinary research in a team context. Peoples noted that when a company hires a chemical engineer, it knows that it can assign him or her a problem and the chemical engineer will know how to tackle it and solve it. Chemists with a bachelor's degree come with the expectation that they will be supervised.

McCoy asked the panel how it could incorporate some of these ideas into the ACS Committee on Professional Training's requirement for a capstone experience. Peoples and Harwell both said that one approach would be to develop scenarios that industry might face and have students form teams to solve those problems. Palmer added that many companies are now posting such problems online and asking for solutions from the community at large. These could be ideal problems for students to tackle.

Coming back to the title of this session—Is there a need for change?—session chair Emilio Bunel of Argonne National Laboratory asked the panel for their final answer to this question. Two of the panelists replied. Bullard said there was an opportunity for change, an opportunity to make chemistry graduates more competitive in the world. Palmer agreed that there was no need for fundamental change, but added, "I think there's a way to make a much bigger impact with what we're already doing."

5

Final Thoughts and Discussion

"We need to come out of the paradigm that we are providers of information. Information is on the web, it is in books, it is everywhere. It does not make sense to have all of that information in our brains . . . we need the strategic knowledge to be able to use the information that is available everywhere."
Miguel Garcia-Garibay

"It seems an incredible travesty to have students walking in the door who devoutly believe that they want to be science majors, a lot of them chemistry majors, who walk out the door with a non-STEM degree. This is a huge loss for the nation."
Susan Olesik

In the workshop's final session, William Carroll, a vice president at Occidental Chemical Corporation and cochair of the Chemical Sciences Roundtable, moderated a panel discussion among chemistry department chairs to get their insights into the state of undergraduate chemistry education and their views on the types of innovations that had been presented in the previous sessions. Members of the panel included Michael Doyle from the University of Maryland; Miguel Garcia-Garibay from the University of California, Los Angeles; Sarah Green from Michigan Technological University; Susan Olesik from The Ohio State University; Jeffrey Reimer of the University of California, Berkeley; and William Tolman from the University of Minnesota. Carroll started the panel discussion by asking each member to take 5 minutes to react to the things they had heard and put them into the context of their own experiences.

Tolman was impressed by the number of creative approaches from dynamic, enthusiastic, and talented faculty who are striving to improve chemistry education. He noted that these cutting-edge approaches do not always produce easily discernible improvements in student learning, but added that the "enthusiasm and talent applied has to be an improvement." He was also struck by Susan Hixson's comments about the missed opportunities to promulgate these novel approaches beyond their home departments and institutions and agreed with her suggestion that chemistry education papers should be embedded in research journals and at scientific conferences. Spreading the word will be key, he said.

Reimer called the workshop presentations "thrilling and uplifting" and stated that he was looking forward to talking to his faculty colleagues about these novel approaches to chemistry education. He was surprised and pleased to hear the industry panelists were of the opinion that chemistry education needs to be fine-tuned to include multiple intelligences but not completely redone. Reimer emphasized that the chemistry community needs to turn "private empiricism"—individuals pursuing new course design based on intuition and experience—into a legitimate scholarly enterprise based on evidence developing in the chemistry education community.

The importance of student-driven activities stood out as a key point that Olesik took from the presentations. "We have to keep reminding ourselves that we need to be the facilitators and not the doers of this work," she said. Angelica Stacy's presentation was important because it drove home the point of how important the design of appropriate exam material is in terms of influencing how students learn and retain information. Finally, Olesik was impressed with the "incredible power that is starting to assemble and the changes that people want to make in teaching chemistry. The world of the biological sciences, and even of physicists, has been moving faster on these fronts and it is really great that the chemists are assigning themselves to this task at a higher level now."

Change of this magnitude is taking years, said Green, and somehow the community must make change happen more quickly. She was struck by the emphasis on multidisciplinary teams and hands-on problem solving based on real-world issues that engages student creativity and worried that the conservative elements of the teaching enterprise will stifle these kinds of courses. Agreeing with Olesik, she said that assessment and evaluation can be important drivers of change.

Garcia-Garibay also agreed with Olesik's point that students' involvement is key and that students can be important agents who add value to knowledge. He noted that chemistry education had done a reasonable job training future chemistry professionals; however, chemistry education is not doing a good job of encouraging students to continue in the broader scientific field and in conveying the importance of chemistry and its fundamental principles to students in the humanities and social sciences. Garcia-Garibay agreed with several presenters that better assessment is needed for how new approaches to chemistry education affect student performance and on the impact of these innovations across the university. He reiterated the early comment that the community needs to work hard at introducing these innovative approaches to much larger audiences and particularly to younger colleagues and teaching assistants.

Speaking from the perspective of having been involved in higher education for 45 years, Doyle said he was impressed with the era of experimentation and innovation that the workshop's speakers represent. He commented on what appears to be a move to use lower-cost methods of instruction that take education away from the master–apprentice approach that has dominated education for so long, and then posed a series of questions that going forward could serve as food for thought for the community.

- What if the National Science Foundation (NSF) had spent money on multiple textbooks that had emphasized interdisciplinary activities in the 1980s and 1990s instead of investing in individual institutions and initiatives that were coming from those institutions?
- What if the American Chemical Society (ACS) Committee on Professional Training had made interdisciplinary education as its mode of approach instead of the subdisciplinary approach that existed in the 1980s and 1990s?
- What if research as an initiation of students to the potential of understanding problem solving and careers in the sciences had moved from a time period that was the capstone experience of a student to a freshman experience that allowed the freshmen to actually start looking at these things early in their educational experience?

Doyle also wondered if the community has the knowledge to understand which of the many approaches presented at the workshop work best and if the nation has the resources to implement any more than one of these approaches. He posed this last issue as a challenge that the community needs to face going forward.

THE CASE FOR CHANGE

Carroll next asked the panel if the case had been made that chemistry education needs to change. Olesik felt that the case for change in the broad field of science education has been made for some time, given the low retention rate for students who express an interest in pursuing a science career when they first enter college and who would be considered the top students based on entering standardized test scores. "It just seems to be an incredible travesty to me to have students walking in the door who devoutly believe that they want to be science majors, a lot of them chemistry majors, who walk out the door with something that is not a STEM major," she stated. "This is a huge loss for the nation." She added that the innovations she heard at the workshop are "spectacular. It is the institutional structures that are a problem."

Students are driving the need for change, said Green, because they have such an evolving smorgasbord of opportunities in front of them. "If they do not like the way we are teaching in our institution, they go somewhere else," she said. Without change, she added, "they are going to vote with their feet or with their dollars." Doyle agreed with this sentiment and noted a program at the University of Maryland College of Engineering that was started 15 years ago when faculty realized that only 38 percent of entering students were graduating in 5 years. The college instituted a program that matches a cohort of 40 students with one faculty member for 2 years with the result that 68 percent of students now finish their degrees in 5 years. "Personal interaction remains a primary determinant on a student's success," he said.

When Carroll asked if anyone wanted to make the case that change was not needed, Tolman said that he did not want to make the opposite case, but refine it. He said that he had not heard the case that fundamental, large-scale institutional change was necessary, but that teaching methods do need to evolve, which should be a natural part of being an educator. Tolman agreed wholeheartedly with the assessment that the community needs to do a better job educating the nonprofessional about science, but that in his mind the evidence was mixed as to whether there is a lack of trained science professionals that is resulting from the low retention rate.

Garcia-Garibay noted that the panel had not addressed the problem of the cost of education, and that is a major driver of change. The cost of education at a large institution such as his is unsustainable, he said, and the major cost of education is faculty. "We need to rethink the paradigm," he said. "How to engage this very expensive faculty in what is becoming an increasingly important portion of the university enterprise?" Carroll asked if chemistry was ripe for the kind of disruptive innovation that could change the cost structure of education, and Tolman replied that massive open online courses (MOOCs) could be such a disruptive force.

Given that the panelists are all department chairs, Carroll noted, he asked them how they plan to drive change in their departments. Tolman said that his department is trying many of these innovations. "We have online sections. We have a MOOC in our department. We have active learning classes, and in fact we have a whole building filled with active learning classrooms that we use with these methodologies. I'm not saying we should not be doing these things. I'm questioning the need for large-scale institutional change throughout the entire system."

Reimer said that one key is practicing what you preach. "When I talk to my colleagues and tell them that I want to have an active learning classroom, I have to do it myself and show it to be successful. That makes my voice far more effective." Part of showing his efforts to be successful, he said, is making sure to demonstrate quantitatively with good assessment tools that the innovative methods are making a difference. "Credibility is an important tool for driving change," he said. Green added that change requires champions willing to take on the task and be rewarded for their efforts, and that requires changing a university's culture in terms of how it values education versus research.

When Carroll then asked for examples of how the panelists introduced change in their departments, Tolman said that he used a video about accidents at two leading universities to introduce a major initiative on safety culture. Reimer said that he held a meeting at which faculty were allowed to have a democratic dialog about a new curriculum that he guided to a predetermined conclusion. Doyle noted that the decision to make a change in the curriculum is out of his hands because so many students from other departments have a chemistry requirement. "Unless we partner with these other departments, our goal of moving toward interdisciplinary curricula is going to be very difficult," he said.

FLEXIBILITY AND SUBJECT MASTERY

From the industry panel's discussions, Carroll had the impression that industry is still interested in subject mastery and depth, but at the same time is looking for students to have technological flexibility. He also noted that a number of the innovative approaches that were presented at the workshop are using context in combination with traditional educational methods. The question he had for the panel was, "Can we take real-world problems and use those to teach the skills that provide flexibility and motivate students while at the same time get to the same depth and mastery that industry seems to be asking for in our students?"

Garcia-Garibay thought that this was possible, though not in every single instance, but what was important was to teach students about the processes of acquiring information, analyzing information, and then transforming that information into action. "We need to come out of the paradigm that we are providers of information," he explained. "The information is in the Web. It is in the books. It is everywhere. It does not make any sense to have all of that information in our brains. We do not need the depth in terms of that information, but we need the strategic knowledge to be able to use information that is available everywhere."

Olesik said that the work that James Anderson and Scott Auerbach described suggests that it is possible to prepare students to be technologically flexible and have a good grounding in the fundamentals of science. Doyle noted that one problem he sees is that this approach might produce students who know how to learn and assimilate science knowledge but that they will not have learned enough content to do well on standardized tests such as the MCAT or Graduate Record Exam.

Carroll responded by asking, "How do we know that we are actually educating better scientists by doing it in a new way?" Tolman added that the key is assessment, but the problem is that most of research faculty, like him, who also teach are not education experts. "We do not know a lot about assessment. We are told to do it, but we do not really learn about it," Tolman said. Carroll acknowledged that short-term assessment was something that the community was going to need to get better at, but the point he wanted to address was whether 5 years down the road students who have passed through these new programs will be better scientists in the workplace. Olesik replied that this was an easy assessment—the companies and institutions that hire these students will either be happy with our product or not. Based on her experience with students who have had an interdisciplinary, deep science class or an active learning class, she thinks that answer will be yes, these students are as good as or better than those who take traditional classes. The panelists also noted that these innovative methods are also giving students better training in nonscience skills such as writing and presentation.

In a final question for the panel, Carroll asked the department chairs if they thought these innovations could be scaled and implemented outside of the home institution. In Tolman's view, the answer is absolutely yes. What it will take, though, is educating faculty so that they want to do it. Reimer agreed and said that funding organizations such as the NSF and Dreyfus Foundation need to continue incentivizing the adoption of these methods, even at a small level. Green also agreed and noted that at her institution, peer-to-peer tutoring has spread so that all classes at Michigan Tech use it to some degree.

TAKING ACTION

As a final activity to close out the discussion, Carroll asked each panelist to state their opinion on what the chemistry community needs to do to accelerate the adoption of the types of innovations presented at the workshop. Doyle said that the sheer number of students that pass through chemistry courses is so large that it has an overwhelming impact on how departments think about their curricula. Instead, he said the focus should be on identifying the students who really need to know chemistry and focus on educating them. Garcia-Garibay returned to the idea that the community needs to figure out how to offer what is a very desirable product in a more economically viable manner. Doing so will require maximizing the value of the most expensive component of college education, the faculty member, and the monologue lecture is not the way to do that. The community needs to figure out how to use technology to address that problem.

Green followed that comment by saying that the reward system has to change to support faculty with innovative ideas. Universities need to encourage faculty to experiment, to take risks, and to fail, just as they do in their research laboratories. Olesik added to that by suggesting that universities need to support interdisciplinary educational programs and develop the financial structures to do so.

With three suggestions on his list of to-do items, Reimer said that the first thing that must happen is for everyone attending the workshop to become leaders at their respective institutions. Second, he would like to see someone develop the MOOC equivalent of the laboratory experience, an activity that he characterized as an interesting intellectual challenge. Finally, he said that someone needs to study, confront, and solve the problem of adolescents in the classroom. "All of my students are smart, but many do not succeed and to a large extent because they are adolescents," Reimer stated.

Tolman also thought the laboratory experience should become an area of focus, but he was of the opinion that the laboratory experience was the one place where in-person, hands-on instruction would not be replaced by a MOOC. "We should be looking at doing things in our labs that inculcate teamwork, cooperation, safety, and culture, all of those things the industrial people want," he said.

Carroll then turned to the workshop attendees and asked each of them to give a one-sentence idea for action based on the workshop's presentations. Their responses were as follows:

- It is important to remember that NSF has funded 20 years of great work upon which this community should draw.
- Make use of the existing body of research on evaluation and instructional methods instead of reinventing the wheel.
- Have the courage to stop innovative programs that are not working.
- Apply the scientific method to teaching—make hypotheses, test them, determine outcomes, and revise those hypotheses in response to data.
- Ensure that each new innovative approach is assessed thoroughly and individually.
- Put additional resources into authentic assessment of innovative methods of teaching, for without authentic assessment there will never be broad change.
- Focus on retention and scientific literacy as key outcome measures.
- Continue to be creative and continue experimenting.
- These innovative methods have given us the opportunity to enhance the education of the best students who are always going to succeed, but also get the attention of the average student who represents the majority of the population.
- There is still a need to develop a new interdisciplinary general education course that meets the science requirement and satisfies university administrators.
- Increase the emphasis on student-centered approaches, which have been shown to increase retention and student preparation.
- Teaching students to be able to read a newspaper article in a scientifically critical manner is the most important skill for them to master.
- Continue developing new approaches that give students skills in collaboration, speaking, and writing.
- Ensure that students who complete these courses have a clear understanding of the process of science, not just the facts of science.
- Keep in mind that the primary job of education should be to transform students from containers of information to creators of knowledge.
- Remember that there is a broad spectrum of different learning goals, some of which can be served by things like MOOCs, but not all.
- Do not minimize the importance of personal interactions between faculty and students.
- The idea of engaging the students in real-world problems is extremely exciting, but the problem is how to scale that up beyond a small number of students.
- Increase the focus on broad-based adoption of even the simple steps that can be taken to improve learning.
- Identify common outcomes so that the community can accelerate the spread of these innovative ideas.
- Support departments implementing new educational paradigms by hiring their students.
- Develop a system that incentivizes teaching that is similar to the way that the current system incentivizes research.
- Make better use of the cohort of current faculty that are serving in adjunct positions.
- Expose tenured faculty at large research institutions to the problems of science education.
- Tap into the larger scientific community outside of the ACS for help in developing principles, strategies, and leadership.
- Include community colleges in this discussion.
- Be sensitive to and aware of the major demographic shifts that have occurred over the past 20 years and that are continuing to change.

GENERAL OBSERVATIONS

In her closing remarks, organizing committee cochair Patricia Thiel summarized some of the key messages that she is taking home from the workshop. Very broadly, she said, it is a great time to be thinking about and talking about education renovation because of two things—great technology and a great foundation of scientific information about what works and what does not work in science education and in chemistry education upon which to carry out renovation. She applauded the efforts to focus educational efforts on global problems and the desire to produce more scientifically literate students. She noted the emphasis on engendering teamwork among students and ownership by students, as well as the importance of designing exams that match the desired outcome goals. She also acknowledged how much work and support are needed for innovations to take hold in institutions. Toward that end, Thiel reiterated the need for innovators to generate evidence that their courses work, that they improve outcomes or maintain outcomes with fewer resources, and that they meet their objectives. "Assessment is important because it will help to convince other people that change is worthy," she said. She also said that it is clear that the community needs to do a better job disseminating these new methods.

She then challenged everyone to think back to the filters they brought to the workshop, to the preconceived biases, and throw them away. "Try to digest the information that was presented in the workshop perhaps through somebody else's point of view," she said. "If you are an educator like me, try to digest them through the point of view of someone who might be funding these programs at NSF or digest them through the point of view of someone who has devoted their lifetime to studying science education and doing assessment."

References

AAMC (Association of American Medical Colleges). 2011. MCAT[2015] A Better Test for Tomorrow's Doctors: Preview Guide for MCAT[2015], November 2011. Available: http://lewisassoc.com/handouts/2015previewguide.pdf (accessed October 17, 2013).

AAMC/HHMI (Association of American Medical Colleges and Howard Hughes Medical Institute). 2009. *Scientific Foundations for Future Physicians*. Washington, DC: AAMC.

ACS (American Chemical Society). 2008. *Undergraduate Professional Education in Chemistry: ACS Guidelines and Evaluation Procedures for Bachelor's Degree Programs*. Washington, DC: ACS.

Brenner, C. 2013. Rethinking premedical and health professional curricula in light of MCAT 2015. *Journal of Chemical Education* 90(7):807-812.

Brenner, C., and D. Ringe. 2012. Response to the new MCAT. *ASBMB Today* (March):12-14.

Brower, A. M. 2009. From teaching to learning. Keynote address to 2009 University of Wisconsin–Madison Teaching & Learning Symposium, May 20-22, 2009 [online]. Available: https://tle.wisc.edu/symposium/archive/2009/keynote-teaching-learning (accessed August 14, 2013).

Cabrera, A. F., and S. M. La Nasa. 2005. Classroom teaching practice: Ten lessons learned. Pp. 129-158 in *Calidad, eficiencia y evaluación de la Educación Superior,* W. de Vries, ed. Madrid: Netbiblo.

Cooper, M. M., N. Grove, S. Underwood, and M. W. Klymkowsky. 2010. Lose in Lewis structures: An investigation of student difficulties in developing representational competence. *Journal of Chemical Education* 87(8):869-874.

Feuer, M. J., L. Towne, and R. J. Shavelson. 2002. Scientific culture and educational research. *Educational Researcher* 31(8):4-14.

Gerlach, K., P. Geissinger, and K. Murphy. 2011. Assessing and enhancing students' scale literacy in introductory chemistry. In *Abstracts of Papers of the 242nd Annual Meeting of the American Chemical Society.* Washington, DC: American Chemical Society.

Gijbels, D., F. Dochy, P. Van den Bossche, and M. Segers. 2005. Effects of problem-based learning: A meta-analysis from the angle of assessment. *Review of Educational Research* 75(1):27-61.

McCoy, A. B., and R. W. Darbeau. 2013. Revision of the ACS Guidelines for Bachelor's Degree Programs. *Journal of Chemical Education* 90(4):398-400.

McNutt, M., ed. 2013. Grand Challenges in Science Education [Special Issue]. *Science* 340(6130):237-396.

NCES (National Opinion Research Center). 1972. National Longitudinal Study of the Class of 1972. ICPSR 8085. Ann Arbor, MI: Inter-university Consortium for Political and Social Research.

NCES. 1980. High School and Beyond, 1980: A Longitudinal Survey of Students in the United States ICPSR 7896. Chicago, IL: National Opinion Research Center.

NCES. 1988. National Education Longitudinal Study of 1988. ICPSR 9389. Chicago, IL: National Opinion Research Center.

NRC (National Research Council). 2000. *How People Learn: Brain, Mind, Experience, and School.* Washington, DC: National Academy Press.

NRC. 2007. *Rising Above the Gathering Storm: Energizing and Employing America for a Brighter Future*. Washington, DC: The National Academies Press.

NRC. 2011a. *A Framework for K-12 Science Education: Practices, Crosscutting Concepts, and Core Ideas.* Washington, DC: The National Academies Press.

NRC. 2011. *Promising Practices in Undergraduate STEM Education.* Washington, DC: The National Academies Press.

NRC. 2012a. *Discipline-Based Education Research: Understanding and Improving Learning in Undergraduate Science and Engineering*. Washington, DC: The National Academies Press.

NPCAST (President's Council of Advisors on Science and Technology). 2010. *Prepare and Inspire: K-12 Education in Science, Technology, Engineering, and Math (STEM) for America's Future*. Available: http://www.whitehouse.gov/sites/default/files/microsites/ostp/pcast-stem-ed-final.pdf (accessed June 7, 2013).

PCAST. 2012a. *Engage to Excel: Producing One Million Additional College Graduates with Degrees in Science, Technology, Engineering, and Mathematics*. Available: http://www.whitehouse.gov/sites/default/files/microsites/ostp/pcast-engage-to-excel-final_2-25-12.pdf (accessed June 5, 2013).

PCAST. 2012b. *Transformation and Opportunity: The Future of the U.S. Research Enterprise*. Available: http://www.whitehouse.gov/sites/default/files/microsites/ostp/pcast_future_research_enterprise_20121130.pdf (accessed June 7, 2013).

Rogers, E. M. 2003. *Diffusion of Innovations*, 5th Ed. New York: Free Press.

Schwartzstein, R. M., G. C. Rosenfeld, R. Hilborn, S. H. Oyewole, and K. Mitchell. 2013. Redesigning the MCAT exam: Balancing multiple perspectives. *Academic Medicine* 88(5):560-567.

Seymour, E., D. Wiese, A. Hunter, and S. M. Daffinrud. 2000. Creating a Better Mousetrap: On-line Student Assessment of their Learning Gains. Paper presented at the National Meeting of the American Chemical Society, March 27, 2000, San Francisco, CA. Available: http://www.salgsite.org/docs/SALGPaperPresentationAtACS.pdf (accessed June 5, 2013).

Shulman, G. I. 2013. Chemistry in the premedical curriculum: Considering the options. *Journal of Chemical Education* 90(7):813-815.

Treisman, P. U., and S. A. Surles. 2001. Systemic reform and minority student high achievement. Pp. 260-280 in *The Right Thing to Do, the Smart Thing to Do: Enhancing Diversity in the Health Professions-Summary on Diversity in Health Professions in Honor of Herbert W. Nickens, M.D.* Washington, DC: National Academy Press.

University of Wisconsin–Madison. 2013. Education Innovation. Available: http://edinnovation.wisc.edu/content/uploads/2013/02/5_step3.pdf (accessed October 17, 2013).

Xie, Y., and A. A. Killewald. 2012. *Is American Science in Decline?* Cambridge, MA: Harvard University Press.

Appendix A

Statement of Task

An ad hoc committee will plan and conduct a public workshop in May 2013 in Washington, D.C. This 1-day workshop will explore the current state of undergraduate chemistry education, with specific consideration of drivers and metrics of change, barriers to implementation of changes, and examples of innovation in the classroom. Research and innovation in undergraduate chemistry education has been done for many years, and one goal of this workshop is to assist in the transfer of lessons learned from the education research community to faculty members whose expertise lies in the field of chemistry rather than in education. This workshop will include a combination of formal presentations and working groups in an effort to stimulate engaging discussion among participants.

The committee will develop the workshop agenda, select and invite speakers and other participants, and moderate the discussions. The focus of the workshop will be on three main goals:

- Identify and examine potential drivers for change,
- Frankly discuss barriers to curricular modifications, and
- Highlight recent results from large-scale innovations with special emphasis on those that are transferrable, widely applicable, and/or proven successful.

Following the workshop, an individually authored summary of the workshop will be prepared in accordance with institutional guidelines.

Appendix B

Workshop Agenda

WEDNESDAY MAY 22, 2013

Welcome and Opening of the Workshop

8:30 Introduction to the Workshop and Its Goals: Patricia A. Thiel, Ames Laboratory and Iowa State University

Drivers and Metrics: Evaluating the Need for Change

8:40 Introduction to the Session by Chair: Miguel Garcia-Garibay, University of California, Los Angeles

8:45 The Role of the ACS Guidelines for Bachelor's Degree Programs: Anne McCoy, The Ohio State University

9:15 Is American Science in Decline?: Alexandra Killewald, Harvard University

9:45 A PCAST Perspective on STEM Education in the New Millennium: S. James Gates, Jr., University of Maryland

10:15 Break

10:45 Chemistry and the Pre-medical Curriculum: The Impact of $MCAT^{2015}$: Joel Shulman, University of Cincinnati

11:15 Lessons Learned at NSF: Susan Hixson, NSF (retired)

11:45 Discussion

12:15 Lunch break

Industry Perspectives: Is There a Need for Change?
Chair: Emilio Bunel, Argonne National Laboratory

12:45 Panelists:
Shannon Bullard, DuPont Chemical Company
David E. Harwell, Assistant Director, Career Management and Development, American Chemical Society
Francine Palmer, Director, Research and Innovation North America, Rhodia/Solvay
Robert Peoples, Executive Director, Carpet America Recovery Effort

1:30 Break

Innovations and Barriers (Part 1)
Chair: Mark Cardillo, Camille and Henry Dreyfus Foundation

2:00 Introduction to the Session: Mark Cardillo

2:15 Propagating Meaningful Reform in Chemistry Education and the Relative Roles of Enthusiasm and Evidence, Thomas Holme, Iowa State University

3:00 Teaching Introductory Chemistry with a Molecular and Global Perspective: The Union of Concepts and Context: James G. Anderson, Harvard University

3:30 Survey of Large-Classroom Reforms: Clark Landis, University of Wisconsin, Madison

4:00 What Gets Measured Is What Gets Learned: Assessing Student Understanding: Angelica Stacy, University of California, Berkeley

4:30 Open Comment and Discussion Period

5:00 Adjourn for the day

THURSDAY MAY 23, 2013

8:30 Welcome to the Second Day of the Workshop: Mark Cardillo, Camille and Henry Dreyfus Foundation

Innovations and Barriers (Part 2)
Chair: Mark Cardillo, Camille and Henry Dreyfus Foundation

8:40 Today's Students and Tomorrow's Leaders: Integrated Concentration in Science: Scott Auerbach, University of Massachusetts, Amherst

9:10 Online Education and MOOCs: Experience with 3.091x, Introduction to Solid-State Chemistry: Michael Cima, Massachusetts Institute of Technology

9:40 Taking Off the Training Wheels: Dealing with Risk, Failure, and Uncertainty: Jeffrey Moore, University of Illinois, Urbana–Champaign

10:20 Break

Wrap-Up Panel Discussion Among Department Chairs
Chair: William Carroll, Occidental Chemical Corporation

10:40 Panelists:
Michael Doyle, University of Maryland
Miguel Garcia-Garibay, University of California, Los Angeles
Sarah A. Green, Michigan Technological University
Susan Olesik, The Ohio State University
Jeffrey Reimer, University of California at Berkeley
William Tolman, University of Minnesota

11:45 Closing Remarks: Patricia A. Thiel, Ames Laboratory and Iowa State University

12:00 Workshop Adjourns

Appendix C

Biographical Information

ORGANIZING COMMITTEE MEMBERS

Emilio Bunel received his Ph.D. in chemistry from the California Institute of Technology in 1988. He began his professional career at DuPont Central Research as a member of the Catalysis Group. He was responsible for the discovery and subsequent development of new processes for the synthesis of Nylon intermediates required in the manufacture of Nylon-6,6 and Nylon-6. In 2001, Bunel was hired by Eli Lilly to establish the Catalysis Group within the Discovery Research Organization. This group was responsible for the preparation of organic compounds using transition metal catalyzed reactions. The molecules prepared spanned all the aspects of the pharmaceutical endeavor from early lead optimization to process development. In 2003, he became an associate director at Amgen, Inc. His work included the establishment of the Catalysis Group in support of route selection/process development efforts to manufacture active pharmaceutical ingredients for clinical testing. Most recently, Dr. Bunel was employed as the director of research at Pfizer, Inc., where he directed the Catalysis Group in support of medicinal chemistry and process development. After spending so many years in industry, Dr. Bunel decided to get back to where science is discovered and not just used. At Argonne National Laboratory, with a talented group of scientists and engineers, but with funding shifting to applied science, he is emphasizing the importance of having a strong basic research program as well.

Mark J. Cardillo is the executive director of the Camille and Henry Dreyfus Foundation. Dr. Cardillo received his bachelor of science degree from Stevens Institute of Technology in 1964 and his Ph.D. degree in chemistry from Cornell University in 1970. He was a research associate at Brown University, a CNR research scientist at the University of Genoa, and a postdoctoral research fellow in the Mechanical Engineering Department at the Massachusetts Institute of Technology. In 1975, Dr. Cardillo joined Bell Laboratories as a member of the technical staff in the Surface Physics Department. He was appointed head of the Chemical Physics Research Department in 1981 and subsequently named head of the Photonics Materials Research Department. Most recently, he held the position of director of Broadband Access Research. Dr. Cardillo is a fellow of the American Physical Society. He has been the Phillips Lecturer at Haverford College and a Langmuir Lecturer of the American Chemical Society. He received the Medard Welch Award of the American Vacuum Society in 1987, the Innovations in Real Materials Award in 1998, and the Pel Associates Award in Applied Polymer Chemistry in 2000.

Miguel Garcia-Garibay has been a faculty member in the Department of Chemistry and Biochemistry since 1992. He came to the University of California, Los Angeles after doing postdoctoral research at Columbia University, which followed his Ph.D. studies at the University of British Columbia, in Canada. The earlier portions of Dr. Garcia-Garibay's education were completed in his native Mexico, at the Universidad Michoacana, where he did research on natural product isolation and characterization. Dr. Garcia-Garibay was promoted to full professor in the year 2000 and he has served as vice chair for education in the Department of Chemistry and Biochemistry since 2005. Dr. Garcia-Garibay is a member of the editorial board of the *Journal of the American Chemical Society* and the *Journal of Organic Chemistry*. He has been a member of the CNSI since 2005. His current research efforts are aimed at the development of artificial molecular machinery in highly organized crystalline media, and to the development of green chemistry by taking advantage of organic reactions in molecular nanocrystals.

Patricia A. Thiel is the John D. Corbett Professor of Chemistry and a Distinguished Professor of Chemistry and of Materials Science & Engineering at Iowa State University. She is also a faculty scientist in the Ames Laboratory. She is active in research, teaching, and administration. In research, she is known for her work in three main areas: nanostructure evolution on surfaces; surface properties and structures of quasi crystals (a complex type of metallic alloy); and the chemistry of water adsorbed on metal surfaces. Dr. Thiel is an enthusiastic teacher of physical chemistry. She has held several administrative posts, including chair of the Department of Chemistry. Dr. Thiel earned her B.A. in chemistry from Macalester College and her Ph.D. in chemistry from the California Institute of Technology in 1981. After postdoctoral work at the University of Munich as a von Humboldt Fellow, she joined the technical staff at Sandia National Laboratories, and then moved to Iowa State University in 1983. In her early academic career, Dr. Thiel was recognized with awards from the Camille and Henry Dreyfus Foundation and the Alfred P. Sloan Foundation, and by a National Science Foundation Presidential Young Investigator Award. Later, she received the American Chemical Society's Arthur W. Adamson Award and the American Physical Society's David J. Adler Lectureship. She was also named fellow of several societies: the American Association for the Advancement of Science, the Materials Research Society, the American Physical Society, and the American Vacuum Society.

SPEAKERS

James G. Anderson is Philip S. Weld Professor in the Departments of Chemistry and Chemical Biology, Earth and Planetary Sciences, and the School of Engineering and Applied Sciences, Harvard University. He was chairman, Department of Chemistry and Chemical Biology, Harvard University, 1998-2001. He was elected to the National Academy of Sciences in 1992, the American Philosophical Society in 1998, the American Academy of Arts and Sciences in 1985, a fellow of the American Association for the Advancement of Science in 1986, and a fellow of the American Geophysical Union in 1989. He is a member of the Space Studies Board of the National Research Council (NRC). The Anderson Research Group addresses three domains at the interface of chemistry and earth sciences: (1) mechanistic links between chemistry, radiation, and dynamics in the atmosphere that control climate; (2) chemical catalysis sustained by free-radical chain reactions that dictate the macroscopic rate of chemical transformation in Earth's stratosphere and troposphere; and (3) chemical reactivity viewed from the microscopic perspective of electron structure, molecular orbitals, and reactivities of radical–radical and radical–molecule systems. He has published over 200 peer-reviewed scientific papers and has testified on numerous occasions for both Senate and House hearings. He was presented the 2012 Smithsonian American Ingenuity Award in the Physical Sciences, the United Nations Environment Programme UNEP/WMO Vienna Convention Award, the Harvard Ledlie Prize for Most Valuable Contribution to Science by a Member of the Harvard Faculty, the ACS National Award for Creative Advances in Environmental Science and Technology, the United Nations Earth Day International Award, the E. O. Lawrence Award in Environmental Science and Technology, the ACS Gustavus John Esselen Award for Chemistry in the Public Interest, the University of Washington Arts and Sciences Distinguished Alumnus Achievement Award, the National Academy of Sciences Arthur L. Day Prize and Lectureship, and the United Nations Environment Programme Ozone Award. He served on the executive committee of the NRC Earth Science Applications from Space: National Imperatives for the Next Decade and Beyond, the Space Science Board Task Group on Research and Analysis; the NRC Committee on Atmospheric Chemistry; NRC Committee on Global Change Research; National Science Foundation Advisory Committee on Atmospheric Sciences; Board of Directors, University Corporation for Atmospheric Research; and Executive Committee and the Pontifical Academy Board for Chemical Events in the Atmosphere and Their Impact on the Environment.

Scott Auerbach is professor of chemistry, adjunct professor of chemical engineering, and founding director of the Integrated Concentration in Science (iCons) program, which focuses on integrating fields of science for training in societal problem areas such as renewable energy and biomedicine, at the University of Massachusetts (UMass), Amherst. He graduated with a Ph.D. in theoretical chemistry from the University of California, Berkeley in 1993 and began his academic position at UMass Amherst in the Chemistry Department in fall 1995. Professor Auerbach won a National Science Foundation Career Award in 1998, a Sloan Fellowship in 1999, and a Camille Dreyfus Teacher-Scholar Award in 1999. In 2006, Professor Auerbach won the UMass College of Science Outstanding Teacher Award. The research of Professor Auerbach and coworkers focuses on advanced materials and catalysts of importance to emerging renewable energy technologies including biofuels and fuel cells, leading to two books and 100 peer-reviewed articles. Professor Auerbach's group also models the molecular-level mechanisms of self-assembly of nanostructured materials.

Michael J. Cima is a professor of materials science and engineering at the Massachusetts Institute of Technology and has an appointment at the David H. Koch Institute for Integrative Cancer Research. He earned a B.S. in chemistry in 1982 (Phi Beta Kappa) and a Ph.D. in chemical engineering in 1986, both from the University of California at Berkeley. Professor Cima joined the MIT faculty in 1986. He was elected a fellow of the American Ceramics Society in 1997. He was elected to the National Academy of Engineering in 2011. He now holds the David H. Koch Chair of Engineering at MIT. He was appointed faculty director of the Lemelson-MIT Program in 2009, which is a program to inspire youth to be inventive and has a nationwide reach. Professor Cima is author or coauthor of over 200 peer-reviewed scientific publications, has 37 U.S. patents, and is a recognized expert in the field of materials processing. He is actively involved in materials and engineered systems for improvement of human health, such as treatments for cancer, metabolic diseases, trauma, and urological disorders. Professor Cima's research concerns advanced forming technology such as for complex macro and micro devices, colloid science, MEMS, and other micro components for medical devices that are used for drug delivery and diagnostics; and high-throughput development methods for formulations of materials and pharmaceutical formulations. He is a co-inventor of MIT's three-dimensional printing process. His research has led to the development of chemically derived epitaxial oxide films for high-temperature superconductivity–coated conductors. He and collaborators have developed a number of drug delivery and diagnostic technologies. Finally, he has been a major contributor to the development of high-throughput systems for discovery of novel crystal forms and formulations of pharmaceuticals. Professor Cima also has extensive entrepreneurial experience as founder and director of several biomedical companies.

S. James Gates, Jr., is the University System of Maryland Regents Professor, John S. Toll Professor of Physics, and director of the Center for String & Particle Theory at the University of Maryland, College Park. He also serves on the U.S. President's Council of Advisors on Science and Technology (PCAST). He has B.S. degrees in mathematics and physics and a Ph.D. degree, all from the Massachusetts Institute of Technology. His thesis, in 1977, was the first at MIT on the topic of supersymmetry. Dr. Gates has held appointments at MIT, Harvard, the California Institute of Technology, Howard University, and Gustavus Adolphus College. He has served as a consultant to the National Science Foundation, the U.S. Department of Energy, the U.S. Department of Defense, the Educational Testing Service and Time-Life Books. Dr. Gates is known for his work in supersymmetry and supergravity, areas closely related to superstring theory, which seeks to describe the fundamental matter of the universe. He authored *Superspace or 1001 Lessons in Supersymmetry* (1984), the first comprehensive book on supersymmetry. He is a past president of the National Society of Black Physicists, and was nominated by Maryland Governor Martin O'Malley to become a member of the Maryland State Board of Education. He is on the board of trustees of Society for Science and the Public. The National Technical Association bequeathed him the National Technical Achiever of the Year and Physicist of the Year awards (1993). The American Physical Society gave him the Bouchet Award (1993). The Washington Academy of Sciences recognized him as the College Teacher of the Year (1999). The University of Maryland has bestowed upon him its Distinguished Scholar-Teacher (2002). The American Association of Physics Teachers presented him with the Klopsteg Award (2003). The National Science Teachers Association recognized him with their Karplus Award (2007). He has appeared in numerous television science documentaries including "The Elegant Universe," "Einstein's Big Idea," "The Fabric of Space," and the BBC's "Hunt for the Higgs." In 2012, he also appeared in the History Channel's "Mankind: The Story of All of Us." Most recently, he has contributed footage for a documentary "The Mystery of Matter," on the development of chemistry. In the last 2 years, Professor Gates has been elected a member of the American Philosophical Society, the American Academy of Arts and Sciences, and most recently, the National Academy of Sciences. During a White House ceremony in 2013, he was bestowed by President Obama with the National Medal of Science, the highest recognition the United States gives in the sciences. The citation on his medal reads, "For contributions to the mathematics of supersymmetry in particle, field, and string theories and extraordinary efforts to engage the public on the beauty and wonder of fundamental physics."

Susan H. Hixson served as a program director in the Division of Undergraduate Education (DUE) within the Directorate for Education and Human Resources at the National Science Foundation (NSF) from 1992 to 2012. Her major responsibilities included serving as the program lead for chemistry within DUE, and as the program lead for the Science, Technology, Engineering, and Mathematics Talent Expansion Program, the Higher Education Centers for Learning and Teaching, the Adaptation and Implementation Track of the Course, Curriculum, and Laboratory Improvement Program, the Systemic Changes in the Undergraduate Chemistry Curriculum Initiative, and the Undergraduate Faculty Enhancement Program. Prior to moving to the NSF, Dr. Hixson was a professor in the Department of Chemistry at Mount Holyoke College from 1973 to 1992, and she also served as chair of the Program in Biochemistry for 6 years during that period. She was a visiting professor in the Department of Biochemistry at the University of North Carolina–Chapel Hill (1980) and a visiting scientist in the Department of Biochemistry and Molecular Biology at the University of Texas Health Science Center in Houston (1986-1987). Her research program at Mount Holyoke focused on the photoaffinity labeling of enzymes with aryl azide reagents. She received her Ph.D. degree in biochemistry from the University of Wisconsin–Madison (1970) and her B.S.

in chemistry from the University of Michigan–Ann Arbor (1965), and served as an instructor at Boston University (1969-1970), and was a postdoctoral fellow at the University of Massachusetts-Amherst (1970-1973).

Thomas Holme is a professor in the College of Liberal Arts and Sciences at Iowa State University. He received his Ph.D. from Rice University in 1987 and was a postdoctoral associate at Hebrew University and the University of Pennsylvania from 1987 to 1989. He began his academic career at the University of South Dakota, coming there from the University of Wisconsin–Milwaukee. He maintains two research groups, one in Chemical Education Research and the other in Computational Chemistry. In Chemical Education, Dr. Holme's research focuses on measurement and assessment of student learning. He serves as director of the Examinations Institute of the American Chemical Society, and his research generally seeks to improve the quality of information that can be obtained from exams and other forms of assessment. The work is carried out within the context of theories of cognition that help organize our understanding of how students approach the tasks they undertake while taking an exam. His group is developing methods to assess the cognitive complexity of test items, considering both the objective complexity inherent in the content covered by the assessment and the subjective complexity as determined by a student taking an exam. The combination of these types of complexity provides an estimation of the cognitive load the student experiences while testing, and this information can help explain the validity and reliability of the measurement of that student's knowledge. In computational chemistry his research group carries out a combination of approaches to look at biologically important chemical processes—in both human and plant applications—that involve chemicals that include main-group inorganic elements such as boron or silicon. Because molecules in this category often contain bonding motifs that have not been extensively studied in biochemical systems, the research begins with small-molecule quantum chemistry studies that inform the development of force-field parameterizations for molecular mechanics calculations.

Alexandra (Sasha) Killewald is an assistant professor of sociology. She received her Ph.D. in public policy and sociology from the University of Michigan in 2011. Her research takes a demographic approach to the study of social stratification. She is coauthor of *Is American Science in Decline?* (2012), which documents trends in the size of the American scientific workforce, public attitudes toward science, youth interest in science, the production of scientific degrees, and transitions to scientific employment.

Clark Landis is professor of inorganic and organic chemistry at the University of Wisconsin–Madison. Born in 1956, he received the B.S. degree in chemistry from the University of Illinois–Urbana and his Ph.D. from the University of Chicago for his work with Jack Halpern on the mechanism of enantioselective hydrogenation. Clark's current research interests center on catalysis and include mechanisms of metal-catalyzed alkene polymerization and enantioselective hydroformylation, development of new nuclear magnetic resonance and mass spectrometric methods for measurement of rapid kinetics, synthesis and applications of modular chiral diazaphospholane ligands, computational modeling of catalytic processes, bonding theory, and chemical education. With Frank Weinhold, he is coauthor of two books, *Valency and Bonding* (Cambridge University Press, 2005) and *Exploring Chemistry with NBOs* (Wiley, 2011). He was the recipient of the American Chemical Society Award in Organometallic Chemistry in 2010 and the University of Wisconsin Chancellor's Distinguished Teaching Award in 2005.

Anne McCoy received her B.S. degree in chemistry from Haverford College in 1987 and her Ph.D. degree in chemistry from the University of Wisconsin–Madison in 1992 and was a Golda Meir postdoctoral fellow with Benny Gerber at the Hebrew University and University of California, Irvine. She joined faculty at The Ohio State University in 1994. She has been a member of the ACS Committee on Professional Training since 2008, served as vice chair of the committee in 2011, and has been the chair since January 2012. She has served as a senior editor for the *Journal of Physical Chemistry* since 2005, and is the deputy editor for the *Journal of Physical Chemistry A*. Professor McCoy has received a number of honors including being named a Camille Dreyfus Teacher/Scholar, giving the Crano Memorial Lecture for the Akron Section of the ACS in 2011, and the Distinguished Scholar Award (Ohio State) in 2013. Professor McCoy is a fellow of the American Physical Society, ACS, and the American Association for the Advancement of Science.

Jeffrey S. Moore received his B.S. in chemistry (1984) and Ph.D. in materials science and engineering with Samuel Stupp (1989), both from the University of Illinois. He then went to Caltech as a National Science Foundation postdoctoral fellow working with Robert Grubbs. In 1990, he joined the chemistry faculty at the University of Michigan in Ann Arbor and then in 1993 returned to the University of Illinois where he is currently professor of materials science and engineering, as well as the Murchison-Mallory Chair in the Department of Chemistry. In 1995, he became a part-time Beckman Institute faculty member under the molecular and electronic nanostructures research theme. He currently serves as lead principal investigator (PI) on four grants including federal (one Multidisciplinary Research Program of the University Research Initiative) and corporate grants. He is also co-PI on four additional grants, working with colleagues across many disciplines. His awards include an Alfred P. Sloan Fellowship and Arthur C. Cope Scholar Award. He has been elected a fellow of the American Association for the Advancement of Science, the Royal Society of Chemistry, the American Academy of Arts & Sciences, and the American Chemical Society. Professor Moore has also received the Campus Award for Excellence in Undergraduate Teaching, the Liberal Arts and Sciences Dean's Award for Excellence in Undergraduate Teaching and has been recognized as a "Faculty Ranked Excellent by Their Students" for his instruction of Chemistry 332. He has served as an associate editor for the *Journal of the American Chemical Society* since July 1999 and advisor of the University of Illinois' Society of Postdoctoral Scholars since January 2011. He has pioneered the development of online organic chemistry courses and is preparing to offer a two-semester organic chemistry sequence as a massive open online course through Coursera. He has over 300 published journal articles covering topics from technology in the classroom to self-healing polymers, mechanoresponsive materials and shape-persistent macrocycles, including publications in *Macromolecules*, the *Journal of Chemical Education*, *Advanced Materials*, and the *Journal of Materials Chemistry*.

Joel Shulman is an adjunct professor of chemistry at the University of Cincinnati. After obtaining a B.S. degree from George Washington University in 1965, he received his Ph.D. in organic chemistry in 1970 from Harvard University. In 1970, he joined the research staff of the Procter & Gamble Company (P&G). During his 31-year career at P&G, he managed projects ranging from drug discovery to the manufacture and commercialization of decaffeinated instant coffee brands to developing ingredients for the first 2-in-1 shampoo. From 1996 to 2001, he was manager of external relations and associate director of corporate research at P&G, with responsibility for bringing new technical capabilities into the company. Included in his department were doctoral recruiting, university relations, external research programs, interactions with government laboratories, and technology acquisition from Russia and China. Upon retiring from P&G in 2001, Dr. Shulman joined the faculty at the University of Cincinnati, where he teaches undergraduate organic chemistry and a course called "Life After Graduate School." He developed this latter course into a 2-day workshop entitled "Preparing for Life After Graduate School," which is presented by the ACS on campuses throughout the country. Dr. Shulman serves the ACS as a career consultant, a consultant to the Graduate and Postdoctoral Scholars Office, chair of the Graduate Education Advisory Board and of the Task Force on the Association of American Colleges and Howard Hughes Medical Institute report, *Scientific Foundations for Future Physicians*, and a member of the Committee on Professional Training. He is a fellow of the ACS.

Angelica Stacy is professor of chemistry and associate vice provost for faculty equity at the University of California (UC), Berkeley. Professor Stacy received her B.A. from LaSalle College in physics and chemistry in 1977 and a Ph.D. in chemistry from Cornell University in 1981. After serving as a postdoctoral fellow at Northwestern University, she began her career at UC Berkeley in 1983. She has published over 120 refereed journal articles, many in the *Journal of the American Chemical Society* and the *Journal of Solid State Chemistry*. She has been a distinguished lecturer at Florida State University (2003), the University of Pittsburgh (2002), and Grinnell College (1999). She received the Catalyst Award from the Chemical Manufacturers Association and the Frances P. Garvin–John M. Olin Medal from the American Chemical Society. In 1984, she received the National Science Foundation's (NSF's) Presidential Young Investigator Award and, in 1991, the Faculty Award for Women Scientists and Engineers. She was cochair of the NSF's Presidential Young Investigators Workshop on U.S. Engineering, Mathematics and Science Education for the Year 2010 and Beyond (1990) and the Gordon Conference on Innovations in the Teaching of College Chemistry (1994). She was an essayist for the Carnegie Project and served on the National Research Council's Chemical Sciences Research Roundtable on Graduate Education. Stacy has received such awards as UC Berkeley's Donald Sterling Noyce Prize for Excellence in Undergraduate Teaching (1996) and the American Chemical Society's James Flack Norris Award for Outstanding Teaching of Chemistry (1998). She also received UC Berkeley's Distinguished Teaching Award in 1991 and was named to the Presidential Chair in Undergraduate Education by UC's Office of the President from 1993 until 1997.

PANELISTS

Shannon Bullard is a human resources and program manager for the DuPont Chemicals & Fluoroproducts Technical organization. She graduated from the University of Delaware with a B.S. in food science and later continued her education obtaining her M.B.A. from Drexel University. Throughout her career at DuPont, she has been involved in leading science and engineering recruiting initiatives for both new college graduates and experienced hires.

Michael P. Doyle received his B.S. degree from the College of St. Thomas in St. Paul, Minnesota, and obtained his Ph.D. degree from Iowa State University. Following a postdoctoral engagement at the University of Illinois at Chicago Circle, he joined the faculty at Hope College in 1968 where he rose to full professor in 6 years and was appointed the first Kenneth Herrick Professor in 1982. In 1984, he moved to Trinity University in San Antonio, Texas, as the Dr. D. R. Semmes Distinguished Professor of Chemistry, and in 1997 he came to Tucson, Arizona, as vice president, and then president, of Research Corporation and professor of chemistry at the University of Arizona. In 2003, he moved to the University of Maryland, College Park, as professor and chair of the Department of Chemistry and Biochemistry. He has received the Manufacturers Association Catalyst Award (1982), the American Chemical Society Award for Research at Undergraduate Institutions (1988), Doctor Honoris Causa from the Russian Academy of Sciences (1994), Alexander von Humboldt Senior Scientist Award (1995), the James Flack Norris Award for Excellence in Undergraduate Education (1995), the Paul G. Gassman Distinguished Service Award (2001), the George C. Pimentel Award for Chemical Education (2002), the Harry and Carol Moser Award (2005), and the Arthur C. Cope Scholar Award (2006). He is a fellow of the American Association for the Advancement of Science, a member of the editorial boards of five journals, and associate editor of *ChemComm*, and an active member of the American Chemical Society. He has written or coauthored 10 books, including *Basic Organic Stereochemistry*, and 20 book chapters, and he is the coauthor of more than 270 journal publications. With 29 years in undergraduate institutions, more than 130 undergraduate students are coauthors of his publications, many with more than two citations, and more than 50 of these coauthors have obtained their Ph.D. or M.D./Ph.D. degrees.

Sarah A. Green is department chair and professor of chemistry at Michigan Technical University. She received her B.A. in chemistry from the University of Minnesota and Ph.D. from the MIT/WHOI Joint Program in Oceanography from the Massachusetts Institute of Technology and Woods Hole Oceanographic Institution. Dr. Green's research focuses on the origin and fate of dissolved organic carbon in terrestrial, lake, and marine environments; methods for detection of free radicals; photochemical transformations of natural and anthropogenic organic compounds in the environment; oxidative degradation reactions; response of aquatic systems to climate change; effects of electrostatic charge and ionic strength on fast reaction kinetics; behavior of metal-contaminated sediments in the Lake Superior basin; fluorescence-based analytical methods; and integration of biological, geological, physical, and chemical data for understanding global cycles. She is a 2013 Jefferson Science Fellow.

David Harwell is the assistant director for Career Management and Development at the ACS. In this role he develops employment and professional development strategies for ACS members and chemical professionals as well as supporting the ACS Committee on Economic and Professional Affairs. Additionally, he provided support to the ACS Presidential Task Force on Innovation and the Chemical Enterprise, and he is the project lead for the Society's new Entrepreneurial Initiatives including the Entrepreneurial Training Program and the Entrepreneurial Resources Center. Before joining the ACS staff, Dr. Harwell obtained his Ph.D. in chemistry at Texas Tech University, worked as a postdoctoral researcher at University of California, Los Angeles, and served on the faculty of the University of Hawaii. In summary, he is a chemist by training and a career counselor by profession.

Susan Olesik is Dow Professor in the Department of Chemistry and Biochemistry at The Ohio State University. She received her B.A. from DePauw University in 1977 and her Ph.D. in 1982 from the University of Wisconsin–Madison, working with James Taylor. She was also a postdoctoral fellow for Milos Novotny at Indiana University from 1982 to 1984 and for Tomas Baer at University of North Carolina–Chapel Hill from 1984 to 1986. She has been a faculty member at The Ohio State University since 1986, being promoted to associate professor in 1992 and professor in 1997. In 1987, she received the American Society for Mass Spectrometry Research Award; in 1990 she received the Eli Lilly Research Award; in 1998 she received a commendation from the National Aeronautics and Space Administration for work on Cassini-Huygen's Probe; and in 2000 she received the AWISCO Woman in Science Award.

Francine Palmer, Ph.D., is the Solvay Research & Innovation director for North America, responsible for the company's R&I Center in Bristol, Pennsylvania. Research at the Bristol laboratory is focused on nanotechnology and advanced materials, organic electronics, and consumer chemicals. Solvay is a Brussels-based international chemical group, strongly committed to sustainable development with a clear focus on innovation and operational excellence. Francine earned a Ph.D. in organic synthesis at the University of Adelaide, Australia, and started her career as a postdoctoral research fellow under Professor Christopher J. Moody at the University of Exeter. In her current role, she is responsible for key competency and talent management and recruitment, as well as being a regional ambassador for academic and government lab institutions and collaborations.

Robert Peoples is the Executive Director and founder of Carpet America Recovery Effort (CARE). In addition, he is also President of the consulting company Environmental Impact Group, Inc. Until August 2012, he was the Director of the ACS Green Chemistry Institute®. In this capacity, he drove the implementation of the principles of green chemistry across the global chemical enterprise. Peoples has been a member of American Chemical Society (ACS) for 35 years, giving him valuable experience and insight into the chemical industry. Immediately prior to becoming Director of the ACS Green Chemistry Institute®, he served as Sustainability Director for the Carpet & Rug Institute. Preceding this position, Bob was Director of Carpet Sustainability and Market

Development at Solutia, Inc., where he was actively involved in carpet recycling and negotiations that led to the formation of CARE and carpet-related health and indoor air quality issues. While there, he helped found the Board of Directors of CARE. Peoples holds a bachelor's degree in mathematics and chemistry from Montclair State University in New Jersey and a Ph.D. in physical organic chemistry from Purdue University. He serves on several local and national boards including the Carpet America Recovery Effort, Georgia Pollution Prevention Advisory Board, and Green Standard.org. He is a member of several organizations including the National Recycling Coalition, Society of Plastics Engineers, and the American Chemical Society.

Jeffrey A. Reimer is the C. Judson King Endowed Professor in Chemical Engineering at the University of California (UC), Berkeley, and a faculty scientist at Lawrence Berkeley National Laboratory. He received his B.S. in chemistry from UC Santa Barbara (1976) and his Ph.D. (1980, chemistry) from Caltech. After 2 years at IBM's Watson Research Laboratory in New York, he joined Berkeley's faculty in 1982. From 2000 to 2005, Reimer was an associate dean in the UC Berkeley Graduate Division where he was responsible for campuswide reviews of doctoral programs; from 2006 until 2011 he was the Warren and Katharine Schlinger Distinguished Professor and chair of Berkeley's Chemical and Biomolecular Engineering Department. In 1998, Professor Reimer won the Donald Sterling Noyce Prize for Excellence in Undergraduate Teaching in the Physical Sciences and was given the AIChE Northern California Section Award for Chemical Engineering Excellence in Academic Teaching. He was awarded the UC Berkeley Distinguished Teaching Award in 2003, the highest award bestowed on faculty for their teaching. Professor Reimer is author or coauthor of over 160 technical papers and reviews, and coauthor (with T. M. Duncan) of the introductory text, *Chemical Engineering Design and Analysis*. Professor Reimer was a Mercator Professor of the Deutsche Forschungsgemeinschaft (DFG) at RWTH Aachen University in 2006. Since that time he has been named a fellow of the American Association for the Advancement of Science and the American Physical Society, and won the 2012 Eastern Analytical Symposium Award for outstanding contributions to magnetic resonance.

William B. Tolman is a Distinguished McKnight University Professor at the University of Minnesota, Twin-Cities. He received a B.S. degree from Wesleyan University, Connecticut, in 1983, and a Ph.D. from the University of California, Berkeley, in 1987. After a postdoctoral period, 1987-1990, at the Massachusetts Institute of Technology, he joined the faculty at the University of Minnesota. He is a member of the Centers for Metals in Biocatalysis and Sustainable Polymers and currently is serving as chair of the Department of Chemistry (since 2009). Among the honors he has received are the Searle Scholars, National Science Foundation National Young Investigator, Camille & Henry Dreyfus Foundation Teacher-Scholar, Alfred P. Sloan Foundation Awards, the Buck-Whitney Medal from the ACS, a research award from the Humboldt Foundation, and a MERIT award from the National Institutes of Health. He is a fellow of the American Association for the Advancement of Science and the ACS. He was associate editor (2007-2012) and is now editor-in-chief of the ACS journal *Inorganic Chemistry*. He served on the Board of Directors of the Minnesota Academy of Sciences from 2009 to 2011, is a member of the Advisory Board of the ACS Petroleum Research Fund and the governing board of the Council for Chemical Research, and served as chair of the Gordon Research Conferences on Inorganic Reaction Mechanisms (2005) and Metals in Biology (2011). Current research in the Tolman group encompasses synthetic bioinorganic and organometallic/polymer chemistry. In the bioinorganic area, the objective is to gain a fundamental structural, spectroscopic, and mechanistic understanding of metalloprotein active sites of biological and environmental importance via the synthesis, characterization, and examination of the reactivity of model complexes. The goal of the Tolman group's research in the organometallic/polymer area is to synthesize and characterize a variety of metal complexes for use as catalysts for the polymerization of cyclic esters. In this collaborative project with Professor M. Hillmyer, particular emphasis is being placed on developing and understanding the mechanism(s) of processes for the controlled synthesis of polymers derived from renewable resources. His work has appeared in more than 175 publications that have been cited more than 10,000 times.

Appendix D

Workshop Attendees

Austin Aluoch
Technical University of Kenya

James Anderson
Harvard University

Scott Auerbach
University of Massachusetts, Amherst

Kevin Belfield
University of Central Florida

Michael Berman
Air Force Office of Scientific Research

Donna Blackmond
Scripps Research Institute

Todd Brethauer
Independent consultant

David Brown
National Science Foundation

Shannon Bullard
DuPont Chemical Company

Emilio Bunel
Argonne National Laboratory

Allison Campbell
Pacific Northwest National Laboratory

Mark Cardillo
Camille and Henry Dreyfus Foundation

William Carroll
Occidental Chemical Corporation

A. Welford Castleman
Pennsylvania State University

Richard Cavanagh
National Institute of Standards and Technology

Michael Cima
Massachusetts Institute of Technology

Paul Craig
Rochester Institute of Technology

Jennifer Curtis
University of Florida

Katherine Denniston
National Science Foundation

Michael Doyle
University of Maryland

Lisa Dysleski
Colorado State University

Joyce Evans
National Science Foundation

Miles Fabian
National Institutes of Health

Ellen Fisher
Colorado State University

Miguel Garcia-Garibay
University of California, Los Angeles

S. James Gates
University of Maryland

Diana Glick
Georgetown University

Joseph Grabowski
National Science Foundation

Sarah Green
Michigan Technological University

David Harwell
American Chemical Society

Susan Hixson
Independent consultant

Thomas Holme
Iowa State University

Jimmy Hwang
Lane College

Mohammad Itani
Georgetown University

Jack Kaye
NASA Earth Science Division

Alexandra Killewald
Harvard University

William Koch
WFK Consulting LLC

John Kozarich
ActivX Biosciences, Inc.

Jay Labov
National Academy of Sciences

Heena Lakhani
National Science Foundation

Clark Landis
University of Wisconsin, Madison

John Leszczynski
University of Maryland

Sandra Loesgen
Oregon State University

Gary Long
Virginia Tech College of Science

Kathy Malone
Triangle Coalition

Michael Mandler
University of Maryland

Luis Martinez
Rollins College

Luigi Marzilli
Louisiana State University

Anne McCoy
The Ohio State University

Michele McLeod
Annenberg Foundation

Steven Metallo
Georgetown University

Bahram Moasser
Georgetown University

Ken Moloy
DuPont Company Experimental Station

Jeff Moore
University of Illinois, Urbana Champaign

Susan Olesik
The Ohio State University

Maria Oliver-Hoyo
National Science Foundation

Francine Palmer
Research and Innovation, Solvay

David Pennington
Baylor University

Robert Peoples
CARE

William Provine
DuPont Chemical Company

Jeffrey Reimer
University of California, Berkeley

Hal Richtol
National Science Foundation

Dawn Rickey
National Science Foundation

Mike Rogers
National Institute of General Medical Sciences

Arlene Russell
University of California, Los Angeles

Mark Schofield
Haverford College

Trevor Sears
Stony Brook University

Joel Shulman
University of Cincinnati

Angelica Stacy
University of California, Berkeley

James Takacs
University of Nebraska–Lincoln

Matthew Tarr
University of New Orleans

Pat Thiel
Ames Laboratory and Iowa State University

William Tolman
University of Minnesota

YuYe Tong
Georgetown University

Jodi Wesemann
American Chemical Society

Ralph Wheeler
Duquesne University

Terry Woodin
National Science Foundation